LEANDER CLASS FRIGATES

HMS LEANDER, as completed, 3/63
Admiralty

LEANDER CLASS FRIGATES

A History of their Design and Development 1958-90

by

Richard Osborne and David Sowdon

DEDICATION

To Ken Purvis and his dedicated team for designing such beautiful and successful frigates.

On 23 April 1974, M. K. Purvis presented a paper, entitled ''Post War RN Frigate and Guided Missile Destroyer Design 1944-69'', to a meeting of the Royal Institution of Naval Architects. During the discussion that followed, D. K. Brown, M Eng, RCNC, FRINA said:

''The author is to be congratulated not only on this most informative paper but also for having designed or at least influenced most of the ships described. In fact, the true title of the paper should be ''Purvis' Navy.''

© **Richard Osborne and David Sowdon 1990**

**Published in 1990 by the World Ship Society,
28 Natland Road, Kendal, LA9 7LT, England**

ISBN 0 905617 56 8

**Companion Volumes: THE HUNTS: John English
THE TOWNS: Arnold Hague**

CONTENTS

Foreword		7
Acknowledgements		7
Figure	British Frigate Development 1945-90	8
Chapter One	The Evolution of the Leander Class Design	9
Chapter Two	LEANDER to ARIADNE	32
Chapter Three	Batch 1 Ikara Conversions	54
Chapter Four	Batch 2 Exocet Conversions	64
Chapter Five	Batch 2A Towed Array Sonar Conversions	74
Chapter Six	Batch 3 Seawolf Conversions	79
Chapter Seven	Overseas Leanders	85
Chapter Eight	The Leanders in Perspective	97
Appendices:	British Leander Class Building Programme	109
	British Leander Class: Conversions	109
	Foreign Leanders and Derivatives	110
	Particulars of British Leander Class Frigates	111
	Machinery Contracts — British Built Ships	112
	British Leander Class: Machinery and VDS Fit	112
	British Leander Class: Weapons and Sensors	113
Sources		117
Index		118
R.I.P.		120

Front cover: H.M.S. LEANDER in July 1979.

ABBREVIATIONS

AA	Anti-Aircraft
AD	Aircraft Direction
ADAWS	Action Data Automation Weapon System
AS	Anti-submarine
ASW	Anti-submarine Warfare
bhp	Brake Horsepower
CAAIS	Computer Assisted Action Information System
CIWS	Close-in Weapon System
CRBFD	Close Range Blind Fire Director
DP	Dual Purpose
ECM	Electronic Countermeasures
ECCM	Electronic Counter Countermeasures
ESM	Electronic Support Measures
EW	Electronic Warfare
FCS	Fire Control System
HF/DF	High Frequency Direction Finding
IFF	Identification Friend or Foe
in	Inch
kg	Kilogram
kW	Kilowatt
MATCH	Manned Torpedo Carrying Helicopter
mm	Millimetre
MRS	Medium Range System
oa	Overall
pp	Perpendicular
shp	Shaft Horsepower
SCOT	Satellite Communications Terminal
STAAG	Stabilised Tachymetric Anti-Aircraft Gun
STWS	Shipboard Torpedo Weapon System
TASS	Towed Array Sonar System
VDS	Variable Depth Sonar
VERTREP	Vertical Replenishment

FOREWORD

HMS LEANDER, which entered service twenty seven years ago, was the first of twenty six ships completed for the Royal Navy during the subsequent ten years. These particularly graceful frigates, which proved to be excellent sea boats that were reliable and popular in service, formed the backbone of the Royal Navy's escort force in the 1970's. Since first entering service, most of the class have been given major reconstructions which restored their fighting power and even today the Leanders form an important component of the British surface fleet.

Furthermore, the design was adopted partially or wholly by the navies of Australia, Chile, Netherlands, India and New Zealand and a total of eighteen ships were built for, or by these nations. The last Indian Leander, VINDHYAGIRI, was completed in 1981, twenty one years after the original design was drawn up. It is likely that this ship will remain in service for up to thirty years, and if so, the design will have achieved a service life of fifty years.

Clearly, the Leanders represent one of the outstanding frigate designs of the post war period, and they were built in greater numbers than any other class of major Royal Navy surface ships since 1945. In this book we have attempted to outline the evolution of these ships and how they have been reconstructed to keep pace with the technological changes which have taken place since 1963.

Richard Osborne & David Sowdon   Nailsea, Avon 1990

ACKNOWLEDGEMENTS

No book of this type can be written without considerable assistance and we are particularly grateful to David Brown R.C.N.C. (former Deputy Chief Naval Architect), David Brown (Head of the Naval Historical Branch), and Dr. Ian Buxton, who have provided so much information and pointed out the errrors in the original text. Michael Crowdy (Founder and Chairman of the World Ship Society) made an enormous contributions to this volume by continually encouraging both authors as well as putting forward many suggestions for improvements in the text. Roger Fry, Richard Lindfield and Brian Stenhouse are thanked for providing valuable information about Leander class refits, Dr David Lush is thanked for the computer graphic work associated with production of the figure showing British post war frigate development while, Andrew Squires is thanked for photographic assistance during the preparation of this volume. With regard to the photographs used to illustrate the text, we are grateful for the assistance of Lieutenant Commander M. H. Larcombe R.N. (Fleet Photographic Officer) and wish to thank T. Bolton, G. Davies (MG photographic), M. R. Dippy, A. Fraccaroli, M. Lennon, Rear Admiral G. P. Niedbalski (Chilean Navy), B. Sullivan, L. van Ginderen and A. Vicary for supplying some of the key photographs used in the book.

Finally, we wish to thank our wives Chris and Hazel for proof reading the manuscript as well as for their patience and encouragement.
R.H.O.   D.A.S.

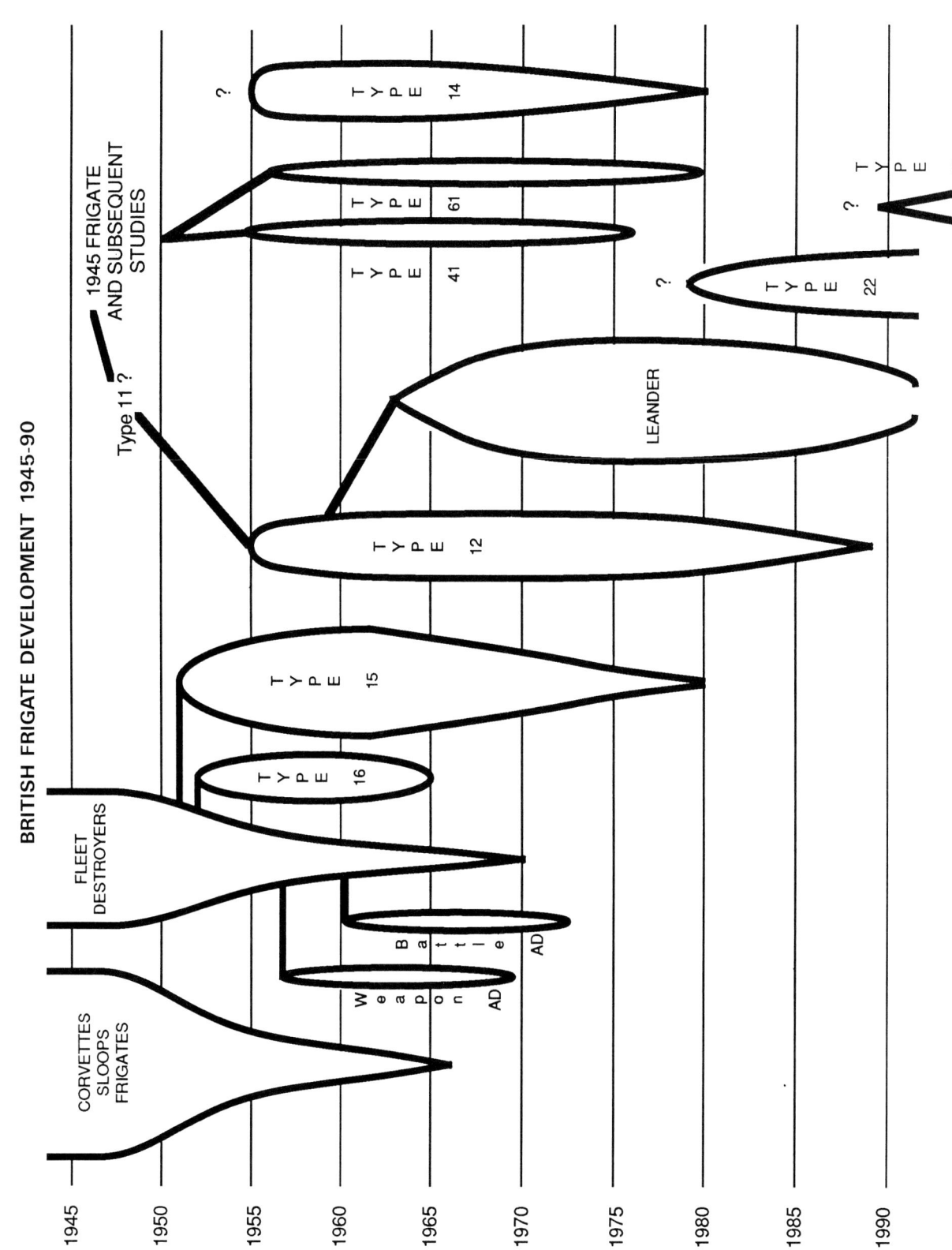

CHAPTER ONE

THE EVOLUTION OF THE LEANDER CLASS DESIGN

There can be little doubt that the British Type 12 and its derivatives, including the Leander class, has proven to be amongst the most successful of post war frigate designs. Thus, at the time of writing, a total of 41 have been built for the Royal Navy with another twenty seven completed for other countries while three 4,000 ton "super" Leanders of the Godavari class have been built for the Indian Navy. Furthermore the Royal Navy has carried the development process through to the highly capable 4,400/4,800 ton Broadsword class frigates.

However, as can be seen from the above, the development of the Broadsword class represents the end result of an evolutionary process that began at least twenty five years before the first of the class was ordered on 8th February 1974. Consequently, in order to understand the origin of the Leander class frigate, it is necessary to consider the development of the Royal Navy's escort forces before and during the Second World War.

DESTROYERS/ESCORTS AVAILABLE FOR SERVICE, POST WAR

Prior to the Second World War the Royal Navy operated two types of escort vessels; destroyers which were intended for fleet duties and sloops which were intended to protect merchant shipping from attack by aircraft and submarines. The need for long-range convoy escorts during the war was so great that hundreds of corvettes and frigates were constructed and these were supplemented by a more limited but nevertheless substantial production of sloops and escort destroyers with good anti-aircraft capability. Soon after the war, all the various types of convoy escorts were reclassified as frigates and construction of fast frigates with a predominantly anti-submarine warfare function continues to this day.

In retrospect it can be seen that the fleet destroyer was obsolete by 1945 because its offensive and defensive functions could be performed far more efficiently by carrier-borne aircraft. Subsequently, the destroyer evolved into fast anti-aircraft warfare escorts and today the traditional destroyer no longer exists. However, for some unknown reason, the Royal Navy continues to categorise AA escorts as destroyers and AS escorts as frigates — an absurd and anachronistic distinction.

In 1945 the Royal Navy possessed some two hundred destroyers most of which had been built during the war. The majority of the war-built ships belonged to the "O" to "Z/C" classes which already were obsolescent and suffered from inadequate gunnery control systems as well as topweight problems. These defects, together with the post-war run-down, condemned the majority of these ships to the Reserve within just a couple of years of the end of the war. By far the most impressive destroyers were the sixteen Early/1942 and eight Later/1943 Battle class ships which were completed in 1944-47. The majority of the Later/1943 Battle class destroyers were cancelled together with all but four Weapon class and all the G class.

Essentially, the Weapon and G class destroyers were designed as fast AA/AS escorts and their cancellation, although inevitable, robbed the Royal Navy of some potentially valuable frigate type ships in the 1950s and 1960s. Production of the Daring class was slowed down and only eight ships were completed from 1952 onwards by which time they were obsolete.

At first sight, the situation with regard to escorts was satisfactory because the Navy possessed several hundred newly completed ships. However, the small Flower class corvettes were discarded rapidly although the larger Castle class were retained for training duties until the late 1950s. The numerous Hunt class escort destroyers were capable of 25 knots but they were cramped vessels with limited endurance and no spare topweight for worthwhile modernisation and consequently, they too were laid up soon after 1945. By comparison, the ships of the Black Swan, River, Loch and Bay classes were armed adequately and were large enough to permit modernisation but were too slow to counter the fast submarines then believed to be under development. The Black Swan class were retained for gunboat style patrol duties in the Middle and Far East while the Loch class were arguably the most efficient and effective anti-submarine ships afloat in the late 1940s.

WAR-BUILT TYPES AVAILABLE FOR POST-WAR SERVICE

HMS ALNWICK CASTLE, Castle class corvette, 8/11/44 — *Admiralty*

HMS MERMAID, Modified Black Swan class frigate, at Trieste 26/3/48 — *R. H. Osborne collection*

HMS VIGILANT, V class destroyer completed in 1943 and obsolescent within five years. Subsequently converted to a Type 15 frigate in the early 1950s
WSPL, Kennedy collection

HMS ANNAN, River class frigate, 1944
Admiralty

HMS LOCH ACHRAY, Loch/Bay class frigate arriving at Portsmouth 7/46 to pay off into Reserve
WSPL, Kennedy collection

HMS ST BRIDES BAY, Loch/Bay class frigate, laid up at Portsmouth 1962, awaiting disposal
WSPL, Kennedy collection

THE 1945 FRIGATE

The British frigates that are in service today are the result of studies carried out by Herbert Pengelly and Ken Purvis, both members of the Royal Corps of Naval Constructors, during 1944-45, to determine the nature of the escort force required for modern warfare. Wartime experience had shown that, because 36 knots provided no defence against aircraft and couldn't be sustained in the mid-Atlantic, the main considerations were ability to keep the sea and robust, but relatively simple construction. In 1944 the Admiralty learned of the high underwater speed of the German Type XXI U-boats and, as existing 19-knot frigates were too slow to counter these fast submarines, formulated a requirement, proposed in January 1945, for a common-hull fast escort with a speed of at least 23 knots.

Two types of frigate were envisaged in the "1945 Frigate Concept", each with a primary role, namely ASW (Anti-Submarine Warfare) and AA (Anti-Aircraft) but with a common hull and steam machinery. It was intended that the hulls would be all-welded and prefabricated in small sections so that they could be assembled very rapidly. Furthermore, because the bulk of the prefabricated units were interchangeable between the two types, a decision, based on the operational need at the time, as to the actual type to be completed, could be delayed until a relatively late stage of construction. These early studies were known as the "25 knot sloop" and were for vessels of 1,650 tons standard displacement (approximately 2,143 tons full load) with a length of 325.5 feet and a beam of 40.5 feet. The ASW version (later Type 12) was intended to be armed with one twin 4.5 in DP gun turret, a twin 40mm Bofors STAAG mounting and two Type D AS mortars, later known as "Limbo", while the AA version (later Type 41) would be equipped with two twin 4.5 in gun turrets, a twin 40 mm STAAG mounting, four single 40 mm guns and a Squid mortar. In February 1946, an AD (Air Direction) version (later Type 61) was added, in which the after 4.5 in gun turret of the AA vessel was replaced by a comprehensive air warning radar outfit.

At about this time, a simple code was drawn up to distinguish between these three streams of escort design:

Type Number	Primary Function of Escort
11 — 40	ASW
41 — 60	AA
61 — 80	AD

and later:

81 —	GP (General Purpose)

With regard to the above code, it has to be emphasised that in the late 1940s the term frigate implied a single-role ship. Multi-role escorts were classified as destroyers if capable of fleet speed, or as sloops if slower. Therefore, the concept of a General Purpose frigate would have been a contradiction in terms at this time.

The Types 41, 61, and 12 groups of ships were designed during the age of the first electronic revolution which was based on thermionic valves, and compared with earlier ships, there was a reduction in weapons and an increase in sensors as designs became volume limited, rather than weight limited. This trend has continued through the age of the transistor and into the microchip era with weapons becoming more voluminous while communications systems, EW equipment and electronic sensors have become ever more complex and sophisticated.

By 1947 it had become apparent that the development of a high speed true submarine was but a matter of time and consequently the speed of the new AS frigates would have to be at least 27 knots, which necessitated the use of steam turbines in the ASW version of the "1945 Frigate". However, in March 1947 the Engineer-in-Chief reported that British industry could not produce the number of high quality steam turbines envisaged and consequently it was decided that the AA and AD frigates, in which speed was less essential, would be diesel-powered. Furthermore, because there would be a considerable delay in providing the necessary steam turbine machinery for the ASW design, the urgent need for fast ASW ships was met by conversion of fast but obsolete destroyers of the O to Z classes. The result was the highly effective and long-serving Type 15 Full Conversion and the less capable Type 16 Limited Conversion.

TYPE 15 FULL CONVERSION

A total of twenty three Type 15 conversions were undertaken by the Royal Navy. The objective of this conversion was to put the latest equipment into a hull that was capable of maintaining high speed in any weather or sea state and still allow the weapons to be used. This involved stripping the destroyer down to her bare hull, extending the forecastle deck aft and building an entirely new superstructure of riveted aluminium containing operations and sonar rooms together with computer control systems, greatly improved radar displays and communications equipment. A new type of enclosed bridge was fitted at the forward end of the superstructure directly ahead of the operations room. The captain was provided with a periscope to permit him to con the ship from the operations room and even the lookouts were given transparent plastic domes in the bridge wings. As converted, the Type 15 frigates displaced approximately 2,800 tons and were capable of 31 knots at full load. Their armament was light consisting of a twin Mk 5 Bofors 40 mm gun mounting forward, a dual-purpose Mk 19 twin 4 in gun mounting aft together with either two Squid or two Mk 10 AS mortars en echelon aft.

Some ships, including the prototype RELENTLESS, were fitted with torpedo tubes for AS torpedoes but the latter proved ineffective and the tubes were soon removed. The radar outfit included a Type 293Q target acquisition radar and a Type 974 navigation radar on the lattice foremast as well as a Type 277Q aerial above the bridge. The Type 293Q radar provided target information for the Close Range Blind Fire director (CRBFD) aft which was fitted with Type 262 radar and controlled the fire of the gun armament. The Type 277Q set was fitted primarily as a surface-warning set to detect surfaced submarines or their "snort" tubes when submerged. A HF/DF aerial was fitted at the head of the foremast which, together with the smaller lattice mainmast, carried a considerable array of communications equipment. As befitted AS ships, the Type 15s were equipped with the most up-to-date sonar sets available including Types 170 and 174.

These conversions took two or more years to complete and were carried out over a period of seven years (1951-57). Initially the Type 15s were regarded with disgust because of their novel appearance and lack of guns. However, later it was realised that they were amongst the best reconstructions ever carried out and provided the Royal Navy with a considerable number of modern fast AS ships at a critical time. Although intended as a stop-gap measure, the Type 15s proved to be successful, popular and durable ships, the last of the class — GRENVILLE — surviving until the mid 1980s.

HMS GRENVILLE, Type 15 frigate 8/4/54 *Admiralty*

HMS TEAZER, Type 16 frigate   WSPL, Kennedy collection

TYPE 16 LIMITED CONVERSIONS

In the early 1950s the Royal Navy carried out ten Type 16 Limited Conversions. The aim was to produce a fast, useful AS unit more quickly and more cheaply than would be achieved by undertaking a full Type 15 reconstruction. The Type 16 conversion was limited to re-arming the ship and fitting as much new equipment as was practically possible in an unmodified hull/superstructure. On completion, these frigates, which displaced some 2,500 tons, were still very much destroyers in appearance and they even retained the after set of quadruple tubes for 21 in torpedoes. The gun armament was reduced to a twin 4 in mounting in "B" position, a twin 40 mm gun mounting amidships and up to five single 40 mm guns. The radar outfit consisted of Types 293Q and 974 on the lattice foremast plus 262 on the CRBF director on the bridge. For AS warfare, these ships were equipped with two Squid AS mortars in "X" position and Type 147P sonar.

The effect of the Type 16 conversion was to produce a "Utility" frigate with an AS armament which was inferior to that of the Type 15. However, the Type 16 had a superior AA and surface armament. The retention of the conventional anti-surface ship torpedo tubes in the Type 16 may have reflected wartime experience with surface commerce raiders and concern at the size of the Soviet cruiser fleet in the mid 1950s. Unlike their fully converted counterparts, the Type 16's spent most of their post-refit lives laid up in Reserve and were disposed of during 1964-68. Plans for Type 18 AS (similar to the Type 16) and Type 62 AD (utilising the five M class hulls) conversions were abandoned in the mid 1950s.

C CLASS DESTROYERS

The widespread use of the CH and CO class destroyers in the post war fleet can be attributed partly to their relative youth and partly to their possession of the Mk 6 director, equipped with Type 275 radar, which improved dramatically their gunnery performance compared with that of the earlier War Emergency Programme destroyers. In the early 1950s, the CH and CO class ships were refitted with two Squid AS mortars in "X" position which improved their AS value. During 1955-59, the eight CA class destroyers were modernised to an even higher standard and were retained in the fleet until the mid to late 1960s and beyond. Two ships, CAPRICE and CAVALIER, were even fitted with the Seacat close range AA guided missile system and consequently remained in service until 1972. Thus, it can be seen that most of the C class destroyers in service by the mid 1950s were more akin to Type 16 frigates than conventional fleet destroyers.

HMS CONCORD, CO class destroyer; the 4.5 in gun mounting originally fitted in "X" position has been replaced by two Squid AS mortars   WSPL, Kennedy collection

HMS SCORPION, Weapon class destroyer, 13/6/53, still in original configuration *Admiralty*

WEAPON CLASS DESTROYERS

In many ways, the Weapon class can be regarded as the forerunners of the modern fast, multi-purpose escorts in service today. For example, they possessed a well-controlled AA armament, two Squid AS mortars (one Mk 10 mortar in SCORPION) and ten 21 in torpedo tubes for anti-surface ship duties. All four ships were taken in hand for modernisation in the late 1950s and emerged as air direction frigates (although officially rated as destroyers). In addition to other changes, both banks of torpedo tubes were replaced by deckhouses and a Type 965 long range air warning radar was fitted at the head of a large lattice mast erected between the funnels. In this new guise, the Weapons served just a few commissions before being laid up for disposal in 1962-64 and scrapped a few years later having been replaced by the more capable Later Battle class conversions.

HMS SCORPION, January 1960 after conversion to an AD ship *Admiralty*

LATER BATTLE CLASS DESTROYERS

Four of this class, AGINCOURT, AISNE, BARROSA and CORUNNA, underwent a similar conversion to that carried out in the Weapon class. Thus, deckhouses replaced torpedo tubes, a launcher for Seacat missiles replaced the 40 mm AA gun armament and a Type 965 air warning radar with AKE-2 array was mounted at the head of a new and massive lattice foremast. The four ships entered service during 1961-63 and remained in operational service for just five or six years before paying off for disposal.

HMS BARROSA, Battle class, after conversion to an AD destroyer   *Admiralty*

POST WAR DESIGNS

LEOPARD (TYPE 41 AA) AND SALISBURY (TYPE 61 AD) CLASS FRIGATES

The drawings for both classes were approved in September 1950 and the first of four ships of each class were ordered in mid-1951. The lead ship of each class was completed in early 1957 and the remaining ships commissioned within a couple of years. They had an unusual appearance with their diesel exhausts taken up into the lattice masts and their distinctive broken-nose profile. Their greatly enlarged, all-welded hull with its raised forecastle markedly improved seaworthiness, as well as providing much needed increased internal volume, and was to be a feature of most succeeding British frigate designs. The main propulsion machinery consisted of eight 16-cylinder ASR1 (Admiralty Standard Range Mk 1) diesel engines, four on each shaft, developing a total of 14,400 bhp, to give a top speed of only 24 knots but a high endurance on fuel carried compared with steam turbine frigates. These diesel engines, which were constructed in Chatham Dockyard, were among the very few surface ship engines designed and built by the Admiralty. As a consequence of the adoption of diesel engines, these frigates were fitted with a water displacement fuel system similar to that used in submarines. However, in the case of a surface ship the system was used to maintain stability whereas in submarines it avoided pressure-light fuel tanks.

The Leopards, which were intended to provide AA defence for convoys, displaced 2,520 tons fully loaded and were armed with two twin 4.5 in DP Mk 6 gun turrets controlled by a Mk 6 director on the bridge and a CRBF director aft, plus a twin 40 mm Bofors STAAG mounting aft for close range AA defence. On the other hand, the ASW armament was weak and consisted of one Squid triple-barrelled mortar plus its associated sonars. As completed, the radar outfit included a Type 960 air warning aerial on the mainmast, Type 974 navigation and Type 293Q target indicating antennae on the foremast, plus Type 275 "searchlights" on the Mk 6 director and Type 262 on the CRBF director. When refitted in the early 1960s, the STAAG mounting was replaced by a single 40 mm gun, and the radar suite modernised by fitting Types 965M, with AKE-1 array, 993 and 978 in lieu of Types 960, 293Q and 974 respectively.

HMS PUMA, Type 41 AA frigate   *WSPL, Kennedy collection*

HMS CHICHESTER, Type 61 AD frigate as completed — WSPL, Kennedy collection

By comparison, the Salisburys sacrificed the after 4.5 in gun turret for Type 982 air search radar mounted on a greatly enlarged after superstructure. These AD frigates underwent several changes during their service careers, the most important being the provision of the massive Type 965M with AKE-2 double array radar on an enlarged mainmast and plating-in of both masts. As in the Leopards, Type 993 and Type 978 radars replaced the ageing 293Q and 974 respectively, while SALISBURY and LINCOLN were fitted with the Seacat surface-to-air missile system in lieu of the after 40 mm guns.

While these eight diesel-powered frigates had the range and equipment necessary for the convoy escort duties for which they were designed, they were embarrassingly slow when in company with other frigates and aircraft carriers. Despite this handicap, all ships gave sterling service for fifteen to twenty years before being phased-out.

HMS SALISBURY, Type 61 AD frigate 20/6/78 in her final configuration — L. van Ginderen collection

TYPE 12 WHITBY CLASS FRIGATES

The Type 12's were designed to protect convoys from submarine attack and to hunt and destroy submarines. The sketch design was put forward on 21st February 1950 and the machinery ordered on 2nd March. Subsequently, the building drawings were approved on 14th December and the first of the six ships ordered in February 1951 with the expectation that the vessel would be laid down later that year for completion in mid-1953. However, the programme was delayed by the post-war need to construct merchant ships as well as the demands of the Korean war and consequently the lead ship, WHITBY, was not laid down by Cammell Laird until 30th September 1952 and eventually entered service on 10th July 1956.

As these frigates were designed to counter submarines with a high underwater speed considerable emphasis was given to the requirement that they should be able to maintain high speed under bad weather conditions. This was given such a priority that the Type 12's were the first RN warships designed for lesser speed loss in bad weather at the expense of a small drop in speed in a flat calm. As noted above the studies which led up to the hull form adopted, started just after the end of the Second World War because war experience had shown that the addition of 3 or 4 knots on to the best speeds available in rough weather was of far greater value than a fraction of a knot at 29-31 knots in a flat calm. In the event, their calm water speed at 30,000 shp was little less than that attained by O-C class destroyers at 40,000 shp partly because of the greater propulsive efficiency obtained from the use of large diameter lower rpm propellers. Consequently, a hull form was adopted with

HMS EASTBOURNE, Type 12 frigate, 1958 as completed: 1 — twin Mk 6 4.5 in gun mounting; 2 — Mk 6 director; 3 — Type 277Q height finding radar; 4 — Type 293Q target indicating radar; 5 — twin 40 mm STAAG mounting; 6 — two Mk 10 AS mortars; 7 — twin trainable 21 in torpedo tubes to port and starboard; 8 — four fixed 21 in torpedo tubes to port and starboard
Admiralty

greater length to displacement ratio than that of the Type 41/61 frigates, with a long, fine-entry, V-shaped bow section, combined with an exceptionally high freeboard. This original hull form was developed by Ken Purvis, who based it on the bow of the destroyer AMAZON and the stern of the cruiser YORK, and further refined by Neville Holt into the Type 12 with its combination of deep draught with a lightweight hull which forced designers to adopt a form with a high rise of floor. To ensure stability, water ballast tanks were placed under the fuel tanks so that as fuel was consumed water was admitted lower down thereby improving stability in the light condition. The origin of the distinctive Type 12 "hump" was explained by Ken Purvis in 1974:

"Amongst the penalties of a fine entry forward for maintaining high speed in rough weather are some loss of stability (GZ) and space low down forward, necessitating moving the forward gun armament further aft which in turn throws the bridge superstructure aft. The need for a clear forward view from the bridge resulted in the lowering of the 4.5 in mounting and to achieve the necessary freeboard forward the Type 12 "hump" was created. This also provided headroom to accommodate the diesel generators".

The raised forecastle deck, which sloped down steeply to the weather deck just forward of the 4.5 in gun turret, when coupled with the recessed anchor housing and marked flare to the bows, acted to reduce the amount of spray thrown up onto the deck and superstructure. Furthermore, a rounded deck edge was fitted in the Type 41, 61 and 12 frigates to minimise stress concentration at beam-frame connections as well as allowing water to flow off more easily thereby reducing the chance of ice formation, a major problem encountered during World War 2 operations in northern waters. The rounded deck edge would also facilitate easy removal of water contaminated with nuclear fall-out should ships operate in a nuclear battle zone. Wartime experience had indicated that structural weakness in ships of destroyer/frigate type occurred at positions of structural discontinuity especially at the break of the forecastle when positioned near amidships. Consequently, in order to eliminate these weaknesses, most post war British frigates were built to a flush-deck design. In the case of the Type 41, 61 and 12 frigates a small quarterdeck was introduced right aft because of misgivings about working wires and other gear at the stern. While these designs entailed an upper deck break, it was in a position of low longitudinal stress and did not give rise to any structural problems. However, the most important reason for the adoption of an otherwise flush deck was the need for additional internal space linked to an improved reserve of buoyancy and stability as well as bad weather access protection for crew and armament aft.

These 2,560 ton frigates were propelled by two sets of Y.100 geared steam turbines (developed by Yarrow-Admiralty Research Department and English Electric) producing 30,000 shp and driving two shafts to give a top speed of 30 knots. They were fitted with slow running (220 rpm) 12 foot diameter propellers which proved to be more efficient and much quieter than the 10.5 foot diameter propellers running at 350 rpm which had been typical until then. As completed, their armament consisted of a twin 4.5 in gun turret forward and a twin 40 mm

HMS BLACKPOOL, as completed. Note the well containing two Mk 10 mortars; torpedo tubes to port and starboard and the STAAG mounting   *R. H. Osborne collection*

STAAG mounting aft plus two triple-barrelled Mk 10 AS mortars on the quarterdeck. In addition there was provision for twelve tubes (eight fixed to port and starboard abreast the aft superstructure and two trainable tubes abreast the STAAG mounting) for Mk 20(E) "Bidder" 21 in acoustic homing ASW torpedoes. Unfortunately, these weapons proved to be ineffective and the tubes were fitted briefly only in BLACKPOOL, EASTBOURNE, SCARBOROUGH and TENBY. The Whitbys were regarded as prototypes and were fitted with the existing twin 4.5 in Mk 6 DP turret in lieu of the twin 3 in/70 cal Mk 6 AA gun mounting intended for production frigates. In the event, this twin 3 in gun proved to be prone to jamming, cost too much and entered service too late to be incorporated into the Royal Navy's frigate programme. The radar suite provided for the Type 12s was similar to that of the Type 41 frigates including Types 293Q, 277Q, 978, 275 and 262 plus Types 162 (bottom search), 170 (attack) and 174 (medium range) sonars.

Despite initial corrosion problems, the Whitbys were successful in service and highly effective against the submerged diesel-electric submarines of the period. However, they entered service eighteen months after the first nuclear submarine NAUTILUS — a new era in ASW had begun and the Whitbys were ill-equipped to counter a nuclear-powered submarine capable of over 25 knots submerged and immune to the effects of adverse surface weather.

TYPE 14 BLACKWOOD CLASS FRIGATES

The origin of the small Type 14 frigates can be traced back to a study, carried out in 1949, to determine the feasibility of building AS frigates smaller than the Type 12s. Minimum requirements were formulated and the twelve ships of the class were ordered in March 1951 with the first ship being completed in December 1955. The intention was to produce the minimum possible AS ship which was also cheap to build and capable of mass production. The result was a long, lean 1,535 ton frigate powered by one set of Y.100 steam geared turbines developing 15,000 shp and driving a single shaft to give a top speed of 24 knots in most sea states. The gun battery was limited to three single 40 mm guns which was all that could be mounted on such a small displacement hull. In any case, it was assumed that modern fast submarines were most unlikely to attempt to fight on the surface. Despite their minimal gun armament, the Type 14s mounted two Mk 10 AS mortars and would have mounted two twin 21 in torpedo tubes for AS homing torpedoes had the latter proven successful. Clearly, the Type 14s possessed the same AS armament as a Type 12 frigate on only half the displacement and in practice proved to be highly effective ASW ships. No worthwhile modernisation was possible because of the small size of the Type 14s and they were the first of the post-war frigates to be scrapped. The Blackwoods were lively and uncomfortable sea boats and their demise must have been a great relief to their crews. In retrospect, it can be seen that the Type 14s were far too specialised and consequently played little part in the future evolution of the modern British frigate.

HMS PELLEW, Type 14 frigate, 13/7/57 as completed with three single 40 mm guns P. A. Vicary

TYPE 12 ROTHESAY CLASS FRIGATES

In 1955 it was decided to update the original Type 12 Whitby design to take account of improvements in weapons and accommodation. Externally, the Whitby funnel, which had been designed to resist nuclear blast, was remodelled to eliminate smoke interference problems and the aft superstructure modified so that the Seacat missile system could be fitted at a later date. Until then, a single 40 mm gun was shipped in lieu. LONDONDERRY

HMS ROTHESAY, Type 12 frigate as completed in 1960. Note twin Mk V 40 mm gun mounting aft and torpedo tubes
WSPL, Kennedy collection

and YARMOUTH were equipped briefly with twelve tubes for Mk 20 (E) acoustic homing torpedoes but their positions were transposed compared to those on the Whitbys with the twin trainable mountings being mounted forward of the fixed tubes. RHYL was completed with only eight tubes for these ASW torpedoes. Originally, twelve Rothesay class frigates were planned for the Royal Navy but HASTINGS (i), which was ordered in February 1956 became the Royal New Zealand Navy's ship OTAGO, while WEYMOUTH, HASTINGS (ii) and FOWEY were completed as LEANDER, DIDO and AJAX of the succeeding Leander class.

HMS LOWESTOFT, 19/10/61 as completed. Note single 40 mm gun aft and absence of torpedo tubes
Admiralty

Bow view of HMS LOWESTOFT 2/6/70 following modernisation. Note plated foremast and single 20 mm guns to port and starboard, abaft bridge
MOD(N)

Stern view of HMS LOWESTOFT 2/6/70. Note Seacat missile launcher on top of helicopter hangar, flight deck and single Mk 10 AS mortar
MOD(N)

Between 1966 and 1972 all nine Rothesays were given a two year refit which involved the installation of the Seacat missile system and the removal of one of the two Mk 10 AS mortars to make way for the hangar and flight deck for a Wasp helicopter. These alterations together with updating of radar and communications systems greatly increased the value and capability of these ships which served the Royal Navy very well for over twenty years.

The Type 12 design assumed that usually sonar ranges would be so short that a 1,000 yard AS weapon, such as the Mk 10 mortar would exploit them fully. In the event of a longer range sonar contact being obtained, the target would be engaged with acoustic homing torpedoes. However, even as the Type 12s were being built, developments in medium range sonar (such as Type 177) made it possible to detect and track submarines at far greater ranges. Consequently, the Royal Navy developed the MATCH (MAnned Torpedo Carrying Helicopter) concept in which a Westland Wasp light helicopter, when guided from its parent frigate, was used to deliver homing torpedoes at ranges greater than those which could be achieved by a similar weapon fired from a surface ship.

The 1951 frigate programme provided the Royal Navy with a reasonable number of modern escort vessels which fell into three specialist types, namely, anti-submarine, anti-aircraft and aircraft direction frigates. This policy of building single function frigates was being questioned by 1954 on grounds of cost and operational problems. Critics argued that it would be far cheaper to build multi-purpose ships with a mixture of AS, AA and AD capabilities and thereby simplify maintenance and operating costs as well as simplifying the operational problem of grouping the correct ships together at the right time. These considerations resulted in the ordering of the first of seven Type 81 Tribal class general purpose frigates in early 1956.

HMS LOWESTOFT arrives at Portsmouth 10/6/82 wearing the paint scheme typical of ships employed on South Atlantic patrol duties
Mike Lennon

TYPE 81 TRIBAL CLASS FRIGATES

This design commenced in 1954 as a project known as the "Common Purpose Sloop" and the intention was to produce a ship with moderate performance in each of the ASW, AA and AD roles compared with that in the single-role specialised frigates then under construction. Eventually, more realistic definitions of escort types were accepted and, as the UK had promised to allocate seventy frigates to NATO, these ships had to be General Purpose frigates. The building drawings were approved in February 1957, although the first ships had been ordered twelve months earlier. ASHANTI was completed by Yarrows in November 1961 while ZULU, the last of her six sisters, did not enter service until April 1964. In many ways, the Tribals were a curious mixture of old and new. Thus, they introduced the COSAG (COmbined Steam And Gas) turbine propulsion arrangement, developing a total of 20,000 shp on a single shaft to give a top speed of about 25 knots which was insufficient to pursue fast submarines. However, the gas turbine installation did enable the ships to make a "cold start" which could be vital in modern warfare. The Tribals were the first British frigates designed with full air conditioning, all bunk sleeping and cafeteria messing.

Broadside view of HMS GURKHA   *R. H. Osborne collection*

In contrast to their modern machinery, the armament of these 2,700 ton frigates was intended to include two twin Mk 19 4 in gun mountings plus two twin Bofors Mk 5 40 mm guns controlled by a CRBF director. However, the 4 in guns in the original design were replaced by single Mk 5 4.5 in DP guns removed from scrapped Z-C class destroyers, because the heavier shell was more suitable for bombardment purposes. In the event, the two single 4.5 in guns with power ramming and remote power control could together fire 40 rounds per minute which was in excess of the normal rate of fire (32 rpm: 2 x 16) of the twin Mk 6 4.5 in gun mounting. Thus, the two single 4.5 in mountings constituted a better AA armament than a twin Mk 6 mounting as well as weighing less. The Tribals were the first frigates to control their armament with the then new MRS3 director which carried its own Type 903 radar and was fitted above the bridge. These ships should have been fitted with the MRS3 Mk II director but this was cancelled thereby causing an increase in their complement.

In the final design the two twin 40 mm gun mountings were replaced by two quadruple Seacat launchers and their associated directors, to port and starboard, but only ZULU was so completed while her sisters had to make do with two single 40mm guns until they were taken in hand for a long refit. The armament was completed by a Mk 10 AS mortar and by provision of a helicopter with its flight deck and hangar aft of the mortar. The lift to the hangar below was made up from part of the flight deck. The helicopter was a very late addition to the design and the machine envisaged was the very small Fairey Ultra Light. The substitution of the much larger Westland Wasp caused many difficulties but many helicopter pilots still rate the Tribal deck as the best frigate

Opposite: Aerial view of the Type 81 Tribal class frigate HMS GURKHA: 1 — single 4.5 in gun; 2 — single 20 mm gun; 3 — MRS 3 director; 4 — Corvus chaff launcher; 5 — Type 993 target indicating radar; 6 — Seacat missile launcher; 7 — Type 965 AKE-1 radar aerial; 8 — Seacat missile director; 9 — Mk 10 AS mortar in well; 10 — Wasp helicopter on landing pad above hangar; 11 — gantry for Type 199 VDS   *MOD(N)*

flight deck because of its high freeboard and lack of eddies. This innovation went some way towards off-setting the low speed of the Tribals relative to that of a submerged nuclear powered submarine but unfortunately the decision to retain a 4.5 in gun on the quarterdeck resulted in very cramped flight deck and hangar arrangements.

The electronic outfit was comprehensive and included Type 965 long range air search radar at the masthead, a Type 978 navigation antenna on the mast and a Type 993 target indicating radar on the bridge in addition to the usual array of sonars including, for the first time, the new medium range Type 177. No major alterations were effected in the Tribals and the last three ships operational with the Royal Navy paid off for disposal in March 1984.

Port quarter view of HMS GURKHA   *R. H. Osborne collection*

Although the Type 81 frigates were expensive, second rate ships they proved that it was possible to build a moderately effective General Purpose escort on a tonnage similar to that of the specialised frigates then in service. The Royal Navy intended to construct twenty three Type 81s or their derivatives (including a "Command" version to replace SURPRISE and ALERT) but their plans were overtaken by successful reworking of the existing Type 12 design.

MODIFIED TYPE 12 LEANDER FRIGATES

In mid 1958 Ken Purvis and his team began to consider the changes necessary to produce a general purpose frigate from the Type 12 design. The process started with a modification of the Rothesay design for the Royal New Zealand Navy incorporating a helicopter and full air conditioning and evolved eventually into a General Purpose frigate, the Leander. This design succeeded in marrying the helicopter and long range air warning radar of the Tribal class with the armament and rough weather capability of the Rothesay class. However, there was no Staff requirement for such a ship as Ken Purvis explained on 13th April 1978 at a meeting of the Royal Institution of Naval Architects:

"Instead of following the normal path of naval construction, design taking requirements laid down by the Naval Staff as the datum, the Leander concept, cheaper and better, was put forward as an alternative to the Type 81 frigate. What is now regarded as a success story from all points of view was opposed originally as being an unnecessary prolongation of the life of an earlier class (i.e. the Type 12 design). Had it not been for the persistent questioning by the Naval Staff who scented the development, it is unlikely that the idea would have been followed up."

The successful outcome of this work was signalled by the Admiralty's decision in January 1960 to complete three Rothesay class frigates to the new Leander design. The major military changes made to the Rothesay design during its evolution to the Leander design were:
— the provision of a flight deck and hangar for an anti-submarine torpedo carrying Wasp helicopter, which necessitated the removal of one of the two Mk 10 mortars,
— installation of improved sonar systems including Variable Depth Sonar (VDS),
— the provision of Type 965 with AKE-1 array air warning radar,
— improved communications systems.

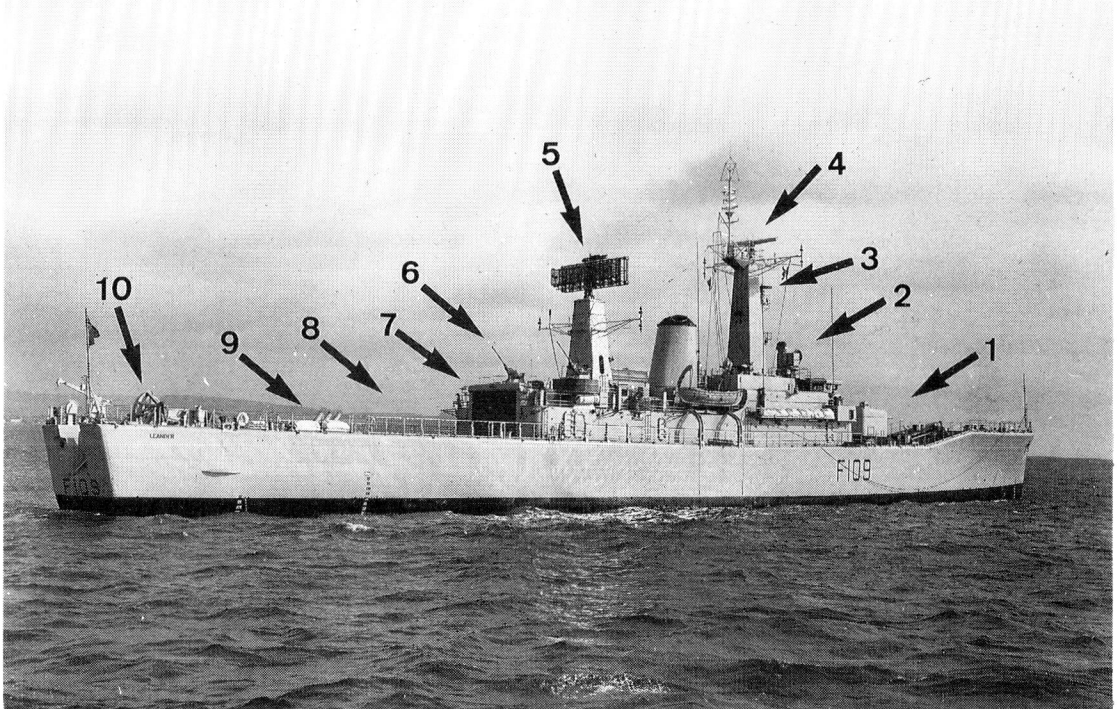

HMS LEANDER, March 1963 as completed: 1 — twin 4.5 in gun mounting; 2 — MRS director; 3 — Type 974 navigation radar; 4 — Type 993 target acquisition radar; 5 — Type 965 AKE-1 air warning radar; 6 — single 40 mm guns port and starboard; 7 — hangar for Westland Wasp helicopter; 8 — flight deck; 9 — single Mk 10 AS mortar in well; 10 — gantry for Type 199 VDS   *Admiralty*

The twin 4.5 in DP gun turret and Seacat missile system of the earlier design were retained. Furthermore, every opportunity was taken to bring the design as up-to-date as possible by incorporating features (detailed below) then being introduced in the County class destroyers as well as providing modern standards of habitability. The latter was achieved by installation of air conditioning equipment which maintained all accommodation and operational compartments at a maximum of 78°F in the tropics and an average temperature of 65°F in cold climates, and the provision of cafeteria messing and bunk sleeping for the entire crew.

Not unnaturally the provision of all these features imposed a considerable space and stability problem for a hull designed originally for a full displacement of 2,560 tons and resulted in some rearrangement of the interior of the ship. Thus, the forecastle deck was extended right to the stern and heavy weights, such as two diesel generators, repositioned as low down as possible, but the bulk of the extra volume required was obtained by elimination of the separate water ballast tanks found in the Whitby and Rothesay classes. Thus, the former water ballast tanks were now used for the stowage of oil fuel by adoption of the oil fuel working, water ballasting and stripping systems which had been developed for the Tribal class frigates and the far heavier County class ships. It was by capitalising on this arrangement that the Leander class was born because as well as providing vital additional stability, the space taken up by separate water ballast tanks in the earlier Type 12 ships could now be used for other purposes. In this new system, sea-water replaced oil fuel in the tanks thereby keeping the centre of gravity low when fuel was consumed.

Weight Breakdown As Designed

Main hull structure	660 tons
General fittings	285 tons
Fittings incidental of armament, machinery etc	210 tons
Electrics and communications	135 tons
Machinery	425 tons
Equipment	245 tons
Armament and ammunition	275 tons
Standard displacement	2235 tons
Reserve feed water	25 tons
Oil fuel	460 tons
Full load displacement	2720 tons

Other features taken from the County class were an improved bridge structure giving greater all-round visibility, especially astern, non-retractable fin ship stabilisers and the installation of hydraulic powered systems to drive boat winches, the after capstan, Seacat missile lifts, helicopter handling gear and the winch controlling the VDS equipment. This greatly reduced the amount of manpower needed on deck. In addition, salt-water jets were provided in the anchor hawse holes to clean cables as the anchors were weighed.

The main hull structural arrangement of the original Type 12 was maintained but was strengthened to cope with the significant increase in displacement from 2,560 to 2,720 tons and design to give a better margin against corrosion. The superstructure was enlarged greatly and a well proportioned funnel sited between two plated masts. The main machinery was essentially that found in the earlier Type 12s but the electrical generating capacity was increased markedly from 1,100 kW to 1,900 kW to meet additional electronic and habitability loads. The capacity of hull and fire pumps was increased to provide an improved pre-wetting capability — the use of seawater spray to wash contamination from the superstructure and decks when operating in NBCD (Nuclear, Biological, Chemical Defence) conditions. As pre-wetting required a pump-capacity of about 200 tons per hour, the pump-capacity was increased to 300 tons per hour.

No portholes were fitted in the hull of the Leander class, as an internal air-tight citadel was provided as defence against nuclear contamination. The operations room, which was incorporated within the citadel, permitted the control and fighting of the ship from a secure position. Access to the citadel was through air-locks fitted with decontamination bathrooms.

Twin propellers and rudders gave excellent manoeuvrability, while the motion of the ship in a seaway, fighting capability and habitability were improved by the installation of non-retractable fin-stabilisers. The pair of stabilisers were fitted at the bilge in line with the mainmast, and initially modifications were needed to the control rods. There was a suggestion that the fins caused additional underwater noise which would impair the performance of the ships' hull-mounted sonars, but tests carried out by Vosper, one of the manufacturers, showed that there was no difference in detection between ships with or without such stabilisers. A marginal decrease in speed did result but was more than outweighed by the overall improvement in fighting capabilities which accrued from the provision of a more stable weapons system platform.

Comparison between the Leander and Tribal designs shows the former to be superior in every respect including sea keeping. Thus, while the Tribals were good sea boats, the Leanders gained rapidly the reputation for being excellent sea boats in all weathers. The respective reputations of the two designs were confirmed in full scale comparative measurements of the behaviour of GURKHA, a Tribal class ship, and HERMIONE, a Leander class vessel, in severe head seas. It was found that the Tribal was unable to complete the run at 20 knots because

HMS LEANDER in heavy weather soon after completion   *Admiralty*

of damage sustained during three heavy seas in succession. In contrast, the Leander suffered no significant damage and was able to complete the 20 knot run and the subsequent 22 knot run in similar weather conditions.

Finally, the Leander design also had the advantage when slamming frequency and perceived motion were considered. Thus, at 16 knots, when the Tribal almost abandoned the run, the Leander's slamming frequency was much less than that experienced on the Tribal. However, HERMIONE seemed to slam at comparatively regular intervals whereas GURKHA would often experience quite long periods of relative calm followed by a quick succession of very heavy slams. Measurements of ships' motions showed that the Leander had only a slight advantage in this respect, but the forward location of the bridge on the Tribal exaggerated the motions experienced by the man in control of the ship.

The Royal Navy's Leander building programme was spread between the laying down of the name ship in April 1959 and the completion of the twenty sixth, ARIADNE, in February 1973. With such a time span many modifications were introduced to individual ships even before the major modernisations began. As completed, the ships fell into three distinct groups: Leander class (10 ships) built 1959-66 with Y.100 steam plant; Phoebe class (6 ships) built 1963-67 with Y.136 steam plant and the Andromeda class (10 ships) built with Y.160 steam plant and two feet more beam, while still maintaining the same compartmentation, to improve stability and provide some space for future modernisation.

The Leanders were described as "General Purpose Frigates", but their design just pre-dated the introduction of guided missiles and long range AS weapons and they were beginning to get outdated by he time the last ship had been built. This had been foreseen a number of years earlier, and in January 1973, LEANDER, the first of the class to be modernised, rejoined the fleet — one month before the last of the class had been completed to a now obsolescent design!

CHAPTER TWO

LEANDER TO ARIADNE

With the design parameters for the Leander class established, the decision was taken in 1960 to alter the last three Rothesay class frigates, then building, to the Leander configuration. This involved increasing the hull by two feet overall, due to the filling-in of the quarterdeck; carrying the forecastle aft; reducing the number of Mk 10 mortars to one and incorporating a considerable number of external and internal changes. The three ships involved were to have been WEYMOUTH, laid down by Harland and Wolff in April 1959; FOWEY, laid down by Cammell Laird in October 1959 and HASTINGS (ii) which had been laid down by Yarrows in December 1959. As these hulls had been laid down only recently, there was relatively little alteration necessary to the existing structure. It was subsequently decided to alter the order for the fifth Type 61 Air Direction frigate to a Leander class vessel. Originally to have been named COVENTRY, the ship was laid down by Vickers-Armstrong at their Walker yard on the Tyne in March 1961.

In 1960 the Government announced the decision to adopt the Canadian version of Variable Depth Sonar (VDS) in preference to the heavier and bulkier British system. The consequence of adopting VDS was that some of the space gained over the Type 12 design by carrying the forecastle aft was lost because of the incorporation of a well for the Type 199 sonar handling equipment in the stern.

The particulars quoted in the Press for the Leander class at this time were: standard displacement 2,235 tons, length 372 feet overall, beam 41 feet and carrying one twin 4.5 in gun turret forward, two 40 mm AA guns aft, one Mk 10 AS mortar and a light helicopter with a hangar. The original displacement for the class was subsequently increased, by the inevitable growth as changes were made, to 2,380 tons standard and 2,860 tons full load. The intention to ship 40 mm guns only as a temporary measure was not fulfilled because the first batch of ships retained them until the early 1970's.

HMS PENELOPE as completed   WSPL, Kennedy collection

In mid 1960 tenders were invited for a further three Leander class frigates and the names of the first four revealed as LEANDER, AJAX, DIDO and PENELOPE respectively. The new tenders were delivered in November 1960 and resulted in contracts being placed with John Brown of Clydebank, Scotts of Greenock and Swan Hunter on the Tyne in 1961. The names for these ships were to be AURORA, EURYALUS and GALATEA respectively. The first, AURORA, was laid down in June 1961, a few days before the launch of the lead ship in Belfast. EURYALUS and GALATEA were laid down in October and December 1961 respectively.

DIDO, the second of the class, was to have been launched on 21st December 1961 but this was prevented by fog. However, the official naming went ahead and the launch took place the following day. In the Spring of 1962 two further orders were placed, one with J. S. White of Cowes for ARETHUSA and the other with Yarrow for NAIAD to be followed by a third order in mid 1962 for CLEOPATRA which was placed with HM Dockyard, Devonport. August 1962 saw the launch of AJAX on the 16th and PENELOPE a day later while Whites launched ARETHUSA on 17th September. That autumn a further three units, each with a modified engine-room layout, were ordered, SIRIUS from HM Dockyard, Portsmouth, PHOEBE from Alexander Stephen of Linthouse and MINERVA from Vickers-Armstrong. In the meantime, NAIAD was laid down on October 30th, 1962 and AURORA launched on 28th November.

New Year 1963 saw LEANDER leave Harland and Wolff's shipyard to commence contractors' sea trials. On completion LEANDER displaced 2,450 tons, an increase over the design figure resulting from minor changes and weight additions. The full load displacement was 2,860 tons, the hull was 372 feet long overall, 360 feet at the waterline, and the beam was 41 feet with a maximum draught at the propellers of 18 feet, but 13.75 feet at the keel amidships.

The engine-room layout was an improvement on that used in the earlier Type 12s, the first ten Leanders being provided with Y.100 machinery arrangement. This consisted of two boilers producing superheated steam at 550 lb/in^2 at 850°F and two double-reduction geared steam turbines giving an output of 30,000 shp. The resulting speed was about 29 knots which was quite sufficient for anti-submarine operations at the time. Control of the engine-room was semi-automatic, thus reducing the number of engine-room staff required.

Boilers for steam turbines can burn a variety of fuels and at an early stage those of the Leander class were modified to use Dieso instead of the heavier Furnace Fuel Oil. There were several main advantages of this — the burning of Dieso caused less damage to the furnace brickwork than FFO; maintenance was reduced; the need for two fuel systems in the Leanders was removed and finally, with the introduction of gas turbines, which had to burn Dieso, it saved the need for the attendant oilers to carry two types of fuel.

Oil-filling arrangements were also improved by the use of a central-filling pipe and distribution system. The oil-filling point on the Leanders was on the foredeck just forward of the superstructure, on which were fitted two replenishment derricks. These permitted refuelling from either beam using replenishment hoses stored aft of the breakwaters to connect with the hose from the oiler. Oil-filling arrangements were dictated by the pump capacity of the supplying ship which was about 600 tons per hour. The tank capacity of a Leander was about 450 tons and so it would take approximately 45 minutes to refuel completely. A system of valves within the ship prevented the tanks being burst by the pump pressure.

The replenishment derricks forward could also be used for taking-on dry stores, in addition to which there was a provision for a light jackstay to be erected on the flight-deck when required. Obviously, the helicopter would need to be either flying or in the hangar for the jackstay to be used. Alternatively, the helicopter itself could be engaged on Vertrep — vertical replenishment operations.

The weapons LEANDER carried when she completed had all been fitted in earlier classes of post-war frigates. Forward of the bridge was the Vickers Mk 6 twin 4.5 in gun mounting, with a weight of 45 tons, which had been introduced as early as 1946. Normally remotely-controlled, although its local control was also possible, loading was semi-automatic. This weapon had a dual-purpose capability although it was intended principally for shore bombardment, the anti-aircraft role being fulfilled more effectively by a smaller calibre weapon. Fire control for these 4.5 in guns was provided by the MRS3 (Medium Range System (Gunnery)) director control tower situated on the bridge top. The system utilised the British Type 903 fire-control radar which was an adaptation of the American Mk 56 fire-control system and its associated Mk 35 radar.

On the hangar roof LEANDER carried, to port and starboard, single 40 mm Bofors AA guns on Mk 9 mountings. It was intended to fit these guns as a temporary measure until such time as the Seacat AA missile system became available for these ships. Seacat had already been fitted in the four County class destroyers which were building at the same time, and in the radar picket conversion of four later Battle class destroyers. In the event, the first seven Leanders were to carry Bofors 40 mm guns until they underwent major refits in the early to mid 1970s.

The ship's anti-submarine capability was provided by the Mk 10 AS mortar, situated in a well aft of the flight-deck. Intended for use against close-range or silent, bottomed targets, the Mk 10 mortar was a three-barrelled system which fired 394 lb (207 lb explosive) mortar bombs in patterns ahead of the ship. To this end the pitch and roll stabilised mounting (weighing 13 tons) was linked through a fire control system to the hull-mounted Type 162 bottom-search and Type 170 attack sonars. The Type 162 was used for classification of targets on the seabed, while the Type 170 "pencil-beam" sonar gave an accurate bearing and depth of the target. With a range of only 1,000 yards and a depth capability of 1,200 ft the weapon was best suited for use against close or quiet submarines against which homing torpedoes would be ineffective. Two salvoes per minute could be fired.

The Leander design concept had evolved around the use of a light helicopter to deliver anti-submarine homing torpedoes and although intended to carry the Westland Wasp helicopter to fulfil this role, none was allocated to LEANDER by the time she was commissioned, and it was not until 1964 that she was so equipped. The Wasp was intended to carry two US Mk 44, later Mk 46, acoustic homing, 12.75 in diameter torpedoes for delivery against targets detected by the new Type 184 medium range sonar, which was being introduced in the Leander class, and the Type 177 sonar. The Mk 44 torpedo had an active homing head and a range of 5 km. In 1964 the Mk 46 torpedo was introduced and featured an active/passive head, increased range, 11 km, and a 10 knot increase in speed to 40 knots. Alternatively the Wasp would be able to carry AS depth charges or two Nord

AS-12 wire-guided anti-ship missiles; additionally the helicopter could be used for reconnaissance, giving the ship an over-the-horizon capability, for light VERTREP, rescue and casualty evacuation — CASEVAC.

Aft of the Mk 10 mortar Leander class frigates carried a new type of sonar — Variable Depth Sonar (VDS) — which was fitted in a ramp cut into the stern and was deployed using a hydraulically operated gantry. VDS had been developed separately by both Canada and Britain from the mid 1950s onwards. The Canadian system underwent initial trials in the destroyer HMCS CRUSADER before a more permanent installation was made in her sister HMCS CRESCENT which had been converted into a Type 15 frigate. Project "Beta" (the code name for the Royal Navy's VDS project, later designated Type 192) began in 1954, and from 1957-60 the Hunt class frigate BROCKLESBY was involved in trials with the heavier and bulkier British system. During one trial, it is

HMS BROCKLESBY, Hunt Type I frigate fitted with the unsuccessful Type 192 VDS which was rejected in favour of the Canadian Type 199 system
WSPL, Kennedy collection

reported that BROCKLESBY almost capsized, due to the transducer hitting a rock, and "Beta" was finally abandoned in 1960. In August of that year it was confirmed that the Canadian system would be adopted by the Royal Navy for its Leander class ships.

Designated Type 199 in British service, VDS was a means of detecting targets in the thermoclines — temperature layers — of the sea. It was possible for submarines to avoid detection by hiding in "shadow zones" afforded by such thermoclines which reflected sonar transmissions. The depth of the thermoclines was established by using the ships' bathythermograph and the operating depth of the VDS controlled accordingly. When the sonar was deployed the gantry or pantograph was lowered from its resting position and the transducer body or "fish" lowered on a cable incorporating the necessary electrical wiring from the ship to the transducer itself. An anti-cavitation fairing was attached to the cable to help reduce noise generated by its passage through the water. Initially it was intended that all Leander class frigates would carry VDS and all but the final pair were launched with a ramp cut in the stern. However, not every ship was equipped with VDS; some only carried the gantry, while others had no equipment at all and eventually had the well plated-in, thereby gaining additional internal space aft.

In service, this sonar did not prove as effective as had been hoped. Its operation required that the ship slow-run, which was far from ideal in the operating conditions prevailing in the Eastern Atlantic, the principal area of Royal Navy deployment in times of tension or hostility. In addition, operating problems occurred with the electrical cables and eventually the decision was taken to fit VDS only in those ships which had been converted to specialist Ikara-armed ASW frigates.

The sonar suite was completed by the Type 185 underwater telephone and the Type 182 towed torpedo decoy. The latter consisted of a noise-generator housed in a towed body, similar in shape to a small torpedo. The decoy was stowed on the port side of the ship, right aft and was deployed by a hydraulic derrick fitted on the same side, between the decoy towing winch and the stowed decoys. Towed astern of the ship, the device was intended to decoy homing torpedoes away from the noise of the ship's propellers.

The radar suite comprised Types 974, 993, 965, 1010 Mk 10 IFF, and the Type 903 fire control set mentioned above. Type 974 (later replaced by Type 978) served as a high definition surface search set and navigation radar. In the first two batches of Leanders this radar was carried on a platform mounted on the front of the foremast, but, in the "Broad-beamed" group and conversions of older ships of the Class the platform was offset to port, on the front of the mast. This change may have been effected to improve detection of vessels aft of the frigate. On top of the foremast was the "quarter cheese" antenna for the then new Type 993 short range air and surface warning radar. First fitted in the Leander class, this radar became one of the Royal Navy's standard target indicating radars and was subsequently retrofitted to earlier classes in lieu of the ageing Type 293Q.

The large "bedstead" AKE-1 aerial for the Type 965 long range air warning radar was carried on the mainmast. The dimensions of this massive aerial, which weighed just under a ton, were 26 feet wide, 8 feet high and 6 feet deep. It consisted of eight elements shaped as reflector horns surrounded by a loop dipole. Atop the Type 965 aerial was the Type 1010 IFF (Identification Friend or Foe) antenna, the two being combined to allow aircraft detected by the Type 965 system to be classified immediately as friendly or otherwise. To complement the active sensors, passive detection was possible with HF/DF and various other devices fitted to the foremast. The increased provision of radar, computers and other electronic equipment, as well as improved living conditions, required an increase in the ships' generating capacity over that of the Type 12 frigates, from 1,100 kW to 1,600 kW in the first batch of Leanders, later increased to 1,900, then 2,500 kW.

Between the masts was the single funnel with the boiler uptakes while the diesel uptakes were incorporated into the foremast, venting aft. Two large ship's boats were fitted, one either side of the funnel, carried in hydraulically operated davits. More general lifesaving gear was provided by a series of liferafts contained in self-inflating canisters, six of which were carried either side of the forward superstructure. A further four liferafts were carried, one either side of the hangar and two at the after end of the hangar roof, in addition to a Gemini inflatable.

In the 1961-62 Naval Estimates the cost of LEANDER was forecast to be £3,750,000 although, by the time she was commissioned on 26th March, 1963 this had risen to £4,810,000. Following her commissioning at Belfast, LEANDER sailed for Portsmouth on 1st April for her first of class trials and work-up. Although not actually allocated a helicopter, the ship took part in deck landing trials with a Wasp during this period. The remainder of 1963 was spent on trials culminating in a first class visit to the West Indies in March 1964. On her return and following a short docking period, the ship became fully operational in August 1964 when she was allocated to the 21st Escort Squadron.

HMS NAIAD, the first of the class to be equipped with Seacat missiles. Note the small director for the GWS 20 Seacat system on the starboard side of the hangar roof
WSPL, Kennedy collection

GALATEA was launched in May 1963, EURYALUS in June, followed by NAIAD and ARETHUSA in November. NAIAD was significant in that she was the two hundredth ship to be built by Yarrow & Co for the Royal Navy in the 98 years of the company's existence, and she was the first of the class to be fitted with a quadruple Seacat launcher aft instead of 40 mm guns. The launcher was mounted on the port side of the hangar aft, with its guidance provided by the GWS 20 director mounted on a low pedestal to starboard. In NAIAD Seacat missile guidance using the GWS 20 system was basically manual in that the missile was tracked through binoculars and directed by the Seacat aimer who used a small joystick.

HMS ARETHUSA as completed. Note the much larger director for the Seacat GWS 22 missile system installed in this and subsequent members of the class
WSPL, Kennedy collection

ARETHUSA was the first ship to be fitted with the more advanced GWS 22 system which was to become the standard director for Seacat. Based on the MRS3 gunnery director, the GWS 22 incorporated the Type 904 radar to track the missile, with an alternative TV director. The GWS 22 director, which was very similar in appearance to MRS3, was mounted on a low platform on ARETHUSA's hangar roof, though in later ships the director was mounted on a higher pedestal. All the Seacat-armed ships were fitted with single 20 mm Oerlikon guns on either side of the superstructure abreast the foremast, for use against small, close-range targets. Although declared obsolete in 1945, the Oerlikon 20 mm gun had been reintroduced as a cheap, simple weapon for use against light targets encountered on patrol duties during the Indonesian confrontation and elsewhere.

In mid 1963, four ships were laid down, CLEOPATRA, PHOEBE — the first of the class with the revised Y.136 machinery layout, and MINERVA and SIRIUS, while tenders were invited for a further three of the class. The order for MINERVA was placed with Vickers-Armstrong who had originally intended to build the ship at their Barrow yard. However, the order was transferred to the company's Tyneside yard.

DIDO, the second Leander, commissioned on 18th September 1963 at the Scotstoun yard of Yarrow & Co and then sailed to Portsmouth. Following trials, DIDO was to have joined the 22nd Escort Squadron in early 1964 but was re-allocated to the 21st Escort Squadron in the Far East. DIDO and PENELOPE, which commissioned on the Tyne at the end of October, each cost £4,600,000 to build, marginally less than LEANDER. AJAX was commissioned in December 1963 at Birkenhead, having been completed at a cost of £4,800,000. Following trials and work-up the ship left Portsmouth on 20th May 1964 to join the 24th Escort Squadron in the Far East as leader. PENELOPE in the meantime had joined the 20th Frigate Squadron at Londonderry.

Early 1963 saw three further orders placed with HM Dockyard, Devonport, Hawthorn Leslie and John I. Thornycroft — to be DANAE, ARGONAUT and JUNO respectively. With sixteen Leanders ordered to date, these three, together with PHOEBE, MINERVA and SIRIUS, completed the six ships of the second batch.

CLEOPATRA, the tenth Leander and the last of the original group with the original machinery layout, was launched on 25th March 1964. The following month GALATEA commissioned on the Tyne before joining AURORA for trials and work-up at Portsmouth. Both were to become squadron leaders, AURORA of the 2nd Frigate Squadron in home waters and GALATEA of the 27th Escort Squadron in the Mediterranean. AURORA ran for a while with an additional gantry fitted on the deck over the VDS well. EURYALUS commissioned in September and became leader of the 26th Escort Squadron at Singapore on arrival on station. Further launches during the year were PHOEBE in July, SIRIUS in September and MINERVA in December.

HMS CLEOPATRA, last of the first batch of ten ships to be completed WSPL, Kennedy collection

On 27th August it was announced that three more ships were to be ordered but of a slightly improved design, in which the beam was to be increased by two feet but retaining the same compartmentation. With new weapon systems being developed the increase gave a greater margin for later modification as well as improving the stability. The first two groups of the class suffered from topweight problems and consequently had to ship permanent ballast to compensate. The new ships would have yet another modified engine-room layout (Y.160) and in addition would be fitted with more fuel tanks, giving an increased endurance. The first three "broad-beamed" Leanders were ordered on 12th January 1965 from HM Dockyard, Portsmouth, Alexander Stephen and Yarrow & Co. at an estimated total cost of £15,000,000. They were to be named ANDROMEDA, HERMIONE and JUPITER respectively.

NAIAD, the first of the Seacat ships, commissioned on 17th March 1965 having cost £4,600,000 to build. At the same time, LEANDER was about to take part in one of the first NATO multi-national exercises and to become Britain's first representative in the NATO "Matchmaker" Squadron. Initially, this consisted of a frigate or destroyer supplied by each of Britain, USA, Canada and the Netherlands and came under the control of the initiator, Admiral Sir Charles Madden, at that time NATO's CINCEASTLANT. The squadron consisted of HMS LEANDER, USS HAMMERBURG, HMCS COLUMBIA and HNLMS OVERIJSSEL and operated together for four months until July 1965, at which time LEANDER went back to the 21st Escort Squadron. The idea proved to be so successful that in January 1968 the Standing Naval Force Atlantic (STANAVFORLANT) was created to provide a continuous patrol by the mixed squadron augmented by ships from West Germany, Norway and Portugal.

In September 1965, following trials and work-up, NAIAD joined the 20th Frigate Squadron at Londonderry, relieving PENELOPE on arrival on station. From this time on PENELOPE was destined for a varied career as a trials ship with the 2nd Frigate Squadron. ARETHUSA completed in September 1965 as the last major ship to be built for the Royal Navy by J. S. White of Cowes, who then announced that their shipbuilding activities were to cease. However, the company's involvement in the Leander programme continued as they built engines for several subsequent orders, and in total contributed machinery to fourteen of the class.

HMS AJAX, 1967, in the Far East. Note deck identification letters (AJ) on the flight deck
R. H. Osborne collection

It is perhaps worth noting the disposition of the Leander class ships at this point in September 1965: LEANDER, 21st Escort Squadron; AJAX, leader of 24th Escort Squadron in the Far East; DIDO, leader of the 21st Escort Squadron in the Far East; PENELOPE, 2nd Frigate Squadron, just prior to trials refit; AURORA, leader 2nd Frigate Squadron on Home Sea Service; EURYALUS, leader 26th Escort Squadron at Singapore; GALATEA, 27th Escort Squadron; ARETHUSA, shortly to join 24th Escort Squadron in the Far East; NAIAD, 20th Escort Squadron at Londonderry; CLEOPATRA, on trials prior to joining 26th Escort Squadron on commissioning and PHOEBE about to commission into the 30th Escort Squadron. MINERVA and SIRIUS were fitting out; JUNO was due to be launched in November 1965; DANAE and ARGONAUT were building, while HERMIONE, ANDROMEDA and JUPITER were on order.

The order for a further two ships of the "Broad beam" design was announced later in 1965 and placed early the following year with Vickers-Armstong on the Tyne, and HM Dockyard, Devonport, the ships becoming BACCHANTE and SCYLLA respectively. In June the twenty-second ship of the class, CHARYBDIS was ordered from Harland and Wolff in Belfast.

Early in 1966 MINERVA started Contractor's sea trials before commissioning on the Tyne on 14th May that year. Meanwhile, PENELOPE started her refit on 1st March in preparation for service as a trials ship with the 2nd Frigate Squadron. This was a role that PENELOPE was destined to perform until paying off for a major refit in December 1977.

HMS PENELOPE following removal of Type 965 AKE-1 radar aerial from mainmast WSPL, Kennedy collection

Initially, PENELOPE had the Type 965 AKE-1 aerial removed from her mainmast, as well as the 40 mm guns and the VDS gear. The 4.5 in gun turret and the MRS3 director were cocooned to protect them from the elements and the VDS well plated-in. During 1968-72 PENELOPE underwent trials with three different designs of "quiet" propellers and as part of this programme skewed blade propellers were fitted for a time. Later, attempts were made to reduce cavitation by the emission of "Agouti" bubbles from the blades of normal propellers. In 1969 it was reported that PENELOPE was to be fitted with a triple torpedo mount on the port side aft — presumably to evaluate the US Mk 32 mount for potential RN service. In 1970 the ship was refitted in preparation for trials aimed at evaluating underwater noise caused by the passage of the ship's hull at varying speeds and conditions. Among the trials conducted at this time was the evaluation of the "Masker" system in which air bubbles were emitted from the underwater hull amidships in an attempt to reduce radiated noise. Such noise adversely affects the detection of submarines while making the ship more of a target. In addition, trials were carried out with Type 184M sonar, a solid state version of the Type 184 fitted in the rest of the class. In due course, this new sonar was fitted into those Leanders modified to carry Exocet missiles, including PENELOPE herself.

In September 1970 while in Gibraltar, PENELOPE had her propellers removed and she was fitted with a pitot rake at the start of underwater noise trials in the Mediterranean. These involved being towed at up to 23 knots by her sister SCYLLA on the end of a one mile long cable, eleven inches in circumference. The trials showed that the hull resistance of PENELOPE was 14 per cent higher than predicted over the trial speed range of 12-23 knots, but that the thrust, torque and efficiency of the ship's propellers were higher than that predicted by model tests.

PENELOPE entered refit yet again during 1973, to be fitted for trials with the Seawolf missile system. All the original armament was removed, as was the MRS3 director. The VDS well was restored but the sonar was not fitted. The mainmast was remodelled slightly to take the back-to-back Type 967/968 radar aerial. Two

HMS PENELOPE photographed in 11/75 while serving as Seawolf missile system trials ship   D. A. Sowdon

deckhouses, containing the electronics associated with the Seawolf system, were fitted on the flight deck. The Type 910 target-tracking radar was mounted on the smaller after deckhouse roof while the sextuple Seawolf launcher was carried on the deck over the position formerly occupied by the Mk 10 mortar well. In this guise PENELOPE carried out sea-trials of the Seawolf system, including tests on the Aberporth missile range. Other trials involved the firing of Seawolf at 4.5 in shells fired by other ships. The ship continued trials until she paid off to commence a major refit at Devonport in December 1977, having completed ten years as a trials ship the previous October.

PHOEBE commissioned in April 1966 as the first of the class with Y.136 modified engine room layout, but was otherwise similar to CLEOPATRA. As completed, PHOEBE was fitted with VDS, although this was subsequently

Stern view of HMS PHOEBE, 6/66, showing gantry and stern well for Type 199 VDS. Note that the VDS "fish" is absent   *MOD(N)*

Port quarter view of HMS PHOEBE, 6/66. Note Seacat launcher to port and Seacat director to starboard on top of hangar roof　　*MOD(N)*

removed and the well plated-in. On commissioning, PHOEBE joined the 30th Escort Squadron. MINERVA commissioned the following month after contractors' sea trials which had begun in January 1966. As with SIRIUS, which had commissioned on 15th June 1966, MINERVA was fitted with a VDS well but did not actually carry the sonar. SIRIUS, which ran for a while without towed decoys or their handling derrick, had her VDS well plated-in by 1970. After acceptance, SIRIUS joined the 24th Escort Squadron, while MINERVA sailed to the Far East.

HMS MINERVA, 1966, entering Portsmouth　　*WSPL, Kennedy collection*

With the fourth ship of this batch, JUNO, was introduced a modified version of the MRS3 gunnery director, though this was not externally apparent. JUNO was built by John I Thornycroft at their Woolston yard and when ordered in early 1964 she gave a much needed boost to a yard which at that time was without a large ship order. Laid down in July 1964 and launched in November the following year, JUNO was also built with a VDS well, but did not carry the sonar. She commissioned at Woolston in July 1967, by which time Thornycrofts had amalgamated with the Portsmouth company, Vosper & Co.

Another yard in need of an order was the Queen's Island Yard of Harland and Wolff in Belfast. Having been mothballed in November 1966, the yard was re-opened for the building of CHARYBDIS, the fifth of the ''Broad-beamed'' Leanders, which was laid down in January 1967. One yard that was not able to survive was the Linthouse yard of Alexander Stephen on the Clyde. While completing PHOEBE the yard had been awarded the contract for the ''Broad-beamed'' ship HERMIONE. Laid down in December 1965, she was launched in April 1967. At the end of that year it was announced that Alexander Stephen was merging with several other yards to form Upper Clyde Shipbuilders. However, it was decided that the Linthouse yard was to be run-down from May 1968 and eventually closed by August 1968. As a consequence, the fitting-out of HERMIONE was completed at the Scotstoun yard of Yarrow, a subsidiary of UCS, with the workforce of the Linthouse yard.

The ''Broad-beamed'' Leander design incorporated several design changes over earlier ships. Thus, the two foot increase in the beam conferred greater stability as well as allowing a greater margin for future alterations while the engine-room arrangement was improved and greater automation introduced. Standard displacement was increased to 2,500 tons but the armament remained the same as that of CLEOPATRA. Although all but the last two of this group were at least launched with a VDS well, the actual sonar and handling gear was fitted only in JUPITER, CHARYBDIS, ANDROMEDA, HERMIONE and BACCHANTE. DIOMEDE was launched with a VDS well, but was plated-in while the ship was fitting out.

Mid 1967 saw the launch of ANDROMEDA, the last warship to be built in HM Dockyard, Portsmouth. At a very early stage ANDROMEDA was fitted with two Corvus eight-barrelled rocket launchers, each surmounted by a 2 in flare projector. These were carried abaft the funnel, to port and starboard on sponsons built on the edge of the shelter deck. Following the sinking of the Israeli destroyer EILAT in 1967 by guided missiles from Egyptian gunboats, it was appreciated that there was a need to fit ships of frigate size and above with some kind of countermeasures to supplement the ECM (Electronic Counter Measures) carried in earlier ships of the class. The Knebworth rockets, fired from the Corvus launchers were designed to release a cloud of aluminium foil (chaff) away from the ship and so decoy incoming missiles. The EILAT incident also showed the need for an effective anti-missile capability as well as demonstrating the destructiveness of the anti-ship missile. The former observation resulted in larger British ships being fitted with Corvus and Leanders already in service were retrofitted; those building were similarly equipped while under construction. Secondly, the effectiveness of the anti-ship missile was to lead to the later modification of some Leanders to take the Exocet surface-to-surface missile.

In the spring of 1967 Yarrow were given an order for the seventh and eighth ''Broad-beamed'' ships which were to become ACHILLES and DIOMEDE. In addition, SCYLLA was laid down at Devonport Dockyard, while JUPITER was launched in September by Yarrow. JUNO commissioned in July 1967, ARGONAUT in August and DANAE, the last of the second batch, in October 1967. On 29th February 1968 BACCHANTE was launched at

HMS DANAE, 1968   *WSPL, Kennedy collection*

HMS CHARYBDIS, 6/76. Note Wasp helicopter landing on R. H. Osborne collection

the Walker yard on the Tyne. The order for the ship had been placed with Vickers in the spring of 1966 but the merger of the Tyne shipbuilding interests of Swan Hunter, Hawthorn Leslie and Vickers-Armstrong meant that BACCHANTE was launched by Swan Hunter and Tyne Shipbuilders. The previous day CHARYBDIS had been launched by Harland and Wolff, but CHARYBDIS was to commission in June 1969, three months ahead of the contract date and almost six months before BACCHANTE. CHARYBDIS was the last warship built by Harland and Wolff until the order for the replenishment ship FORT VICTORIA was placed in 1986.

SCYLLA was launched at Devonport in August 1968 and ACHILLES in November at Yarrow, who had recently been awarded the contract for the last two Royal Navy Leander class ships. This brought the number to be built for the Royal Navy by Yarrow to seven out of twenty six, (plus others for New Zealand and Chile), and the class to the largest of frigate-size and above to be built for the Royal Navy since the end of the Second World War.

The Defence White Paper of 1968, in announcing the final Leander orders, also heralded the decision to modify some of the earlier ships into pure ASW frigates by fitting them with the Australian-designed Ikara AS missile. Thus, before the class had even been completed, some of its members were to be given a major modification. The Ministry of Defence had been involved for some time with the development of the Ikara missile for use in British ships and it had already decided to fit the missile in the Type 82 destroyers then on the drawing board. In the event, however, only one Type 82, BRISTOL, was actually built.

ANDROMEDA, although laid down six months after HERMIONE, was completed in July 1968 but did not commission until December of that year, at which time ANDROMEDA was fitted with VDS, but this was subsequently removed and the well plated-in. The completion of HERMIONE was slightly delayed because of her transfer to the Yarrow yard for fitting-out.

HMS HERMIONE, 6/77 Mike Lennon

ANDROMEDA showed various internal modifications compared to the two earlier batches. Engine-room control had been further automated and was now exercised from a separate machinery control room. New living-area standards were adopted and included separate recreation spaces and multi-berth cabins for CPO's. Further improvements in living standards were officially approved in 1971 and were fitted in all the Leanders to undergo major refit and modernisation. This left such a disparity in the accommodation between the modified ships and the six unconverted Leanders that men serving in JUNO, APOLLO, DIOMEDE, BACCHANTE, ARIADNE and ACHILLES during the late 1970s, were granted additional "hard-lying" money. In the "Broad-beamed" ships the navigation radar platform, which had been fitted on the front face of the foremast previously, was moved 45° and angled out to port, while the earlier Type 974 radar was replaced by Type 978, improvements subsequently made to all the Leanders. The generating capacity was further increased to 2,500 kW to cope with the increased power required for electronic and domestic needs.

The normal complement of a Leander class frigate was 260, made up of 14 officers and 246 ratings. In 1969 an attempt was made to reduce this figure in NAIAD. An arbitrary reduction of 50 was made to assess the effectiveness of such a measure. The trial proved successful and was extended to other ships in the Fleet. Several other navies were by this time operating Leanders with smaller crews compared with those in British ships of the class. The size of crews had been justified in British ships on the grounds that they were maintaining a higher level of readiness for action and so required a larger complement.

On 16th May 1969 three Leander class ships were among the representatives at the NATO 20th Anniversary Review held at Spithead; these were DIDO, at that time the UK representative in the Standing Naval Force Atlantic, PHOEBE and the Dutch-built Leander VAN NES.

Aerial view of HMS PHOEBE, 13/6/69. Note Corvus chaff launchers fitted to port and starboard abreast mainmast
MOD(N)

DIOMEDE, the twenty fourth Leander, was launched by Yarrow on 15th April 1969, while JUPITER and BACCHANTE commissioned in August and December respectively and APOLLO, the penultimate Leander, was laid down in November the same year. Shortly after completing her work-up BACCHANTE was assigned to the NATO Squadron to replace LEANDER in May 1970, an exchange which was the prelude to LEANDER paying-off for reconstruction.

With the impending refit of LEANDER to carry the Ikara missile system, it was announced that in future Devonport Dockyard would undertake the majority of frigate refits. Although it was originally intended to convert twelve ships to carry Ikara this was reduced to eight at an estimated £30,000,000 for the whole programme — clearly, the rampant inflation of the 1970s played havoc with these figures. LEANDER's refit actually cost £7,600,000 while that of DIDO cost £23,000,000!

HMS ACHILLES 21/7/70, one month after completion   MOD(N)

SCYLLA commissioned on 14th February 1970 and after work-up went to the Mediterrean where she assisted PENELOPE in the underwater noise trials. In 1975/6 SCYLLA reputedly ran with a boxed Seawolf missile on deck to assess the effect of the weather on a missile stowed in its launcher. ACHILLES, like SCYLLA and DIOMEDE, was launched with a VDS well cut in the stern, but had this plated-over within a short time. Another Yarrow-built ship, ACHILLES, commissioned on 11th September 1970. The same yard had laid down the final British Leander, ARIADNE, on May Day 1970 and launched APOLLO in the same year. Neither of these ships was ever fitted with a VDS well.

In October 1970 LEANDER paid-off at Devonport to commence a major refit which was to alter radically her appearance and capabilities. The Leanders had been designed before guided missiles came into service and consequently they were threatened with block obsolescence unless brought up-to-date. In the eight years that LEANDER had been in commission great improvements had been made in weapons and sensors and better living conditions were required.

Even before LEANDER's refit was complete, there were seven different armament/sensor outfits within the class, as well as minor variations between ships from different builders. In addition, various modifications were made to ships as they entered dock for short refits and docking periods. All ships of the second and third groups were fitted with Corvus rocket launchers during short refits, had they not been so equipped while building. The majority of the first group gained Corvus, if not already fitted, during their Ikara refits while other improvements included the introduction of the solid state Type 994 search radar instead of the older Type 993, although the same aerial was used in both systems.

Progressively through the early 1970s those ships selected for the Ikara refit paid-off for conversion. Although Devonport Dockyard was responsible for the majority, ARETHUSA refitted at Portsmouth and AURORA at Chatham. In order, the ships converted at Devonport were LEANDER from 1970, followed by AJAX — 1970, GALATEA — 1971, NAIAD and EURYALUS — 1973. ARETHUSA and AURORA started their refits in 1973 and 1972 respectively, while DIDO which had been scheduled to refit at Chatham, ran until 1975 before commencing her modernisation in July.

HMS DIDO, the second Leander to be completed and still in her original configuration when photographed at Wellington on 19/9/73
L. van Ginderen

ANDROMEDA had been in commission for about eighteen months in June 1970, when on patrol in the Indian Ocean, she was called to assist the RFA ENNERDALE which had struck a submerged reef and was sinking. Although ANDROMEDA was able to help the crew, the tanker sank in shallow water and was later destroyed by mortar fire and bombs from naval helicopters.

Spring 1971 saw DIOMEDE completed by Yarrow and on commissioning she became a training ship. While serving in this role her Type 965 radar aerial was removed and her 4.5 in gun cocooned, although both were restored when DIOMEDE became fully operational in due course.

Meanwhile, plans were afoot to restore the firepower of other members of the class by the replacement of the obsolete Mk 6 4.5 in gun mounting by French Exocet surface-to-surface missiles. Exocet had been under development since just before the sinking of the EILAT by Russian-built missiles in October 1967. Following successful French trials, the Royal Navy decided to fit Exocet in four of the County class destroyers and then the remaining Leanders. CLEOPATRA was the first of the class to be taken in hand for this conversion, at Devonport in July 1973. Of the second batch, PHOEBE started her Exocet refit in August 1974, followed by SIRIUS (March 1975) and ARGONAUT (February 1976), all of which were converted at Devonport, while MINERVA paid-off in December 1975 to refit at Chatham. DANAE was scheduled to start her refit at Devonport in June 1976 but this was postponed until August 1977 when she was joined by the Batch I ship PENELOPE for a similar conversion.

Operational commitments and refit schedules were upset in early 1976 with the "Cod War" in the Arctic waters around Iceland. Several Leanders were taken from other duties to act as Fishery Protection ships in the confrontation with Icelandic vessels over disputed fishing rights in those northern waters. In the first instance, the Ikara conversions GALATEA and LEANDER were sent north in December 1975, to be joined by ANDROMEDA and various other frigates. On 28th December, ANDROMEDA was involved in a collision with the gunboat TYR and lost 60 feet of her guardrail, as well as suffering damage to a rocket launcher. A second collision caused a 12 foot dent in ANDROMEDA's port quarter and after this engagement the ship was withdrawn to repair in Devonport.

HMS ANDROMEDA, 5/76. Note "headlamp" antenna at the foremast head. ANDROMEDA was the only non-Ikara armed Leander to be so equipped   *D. A. Sowdon*

During her own return to the UK, LEANDER suffered defects in one of her boilers which had to be shut down. Subsequently, the ship lost all power and steering in a storm 200 miles NW of Scotland — guardrails were bent, and aerials, liferafts and a gangway were lost before the ship, assisted by BACCHANTE, reached Faslane. After temporary repairs, LEANDER sailed for Devonport. JUNO and DIOMEDE suffered rammings and collisions before returning to the UK from the Arctic. JUNO also suffered a small fire aft, as petrol cans in stowage ignited following a collision with TYR. DIOMEDE was hit several times by BALDUR and returned to the UK with a 20 foot gash in her port side, the gunboat's bows having penetrated the hull by about three feet, wrecking the wardroom. NAIAD and SCYLLA were also involved for a short while, NAIAD suffering a gash in her side, followed by a bow damage after another collision. LEANDER, repaired and back in the fray, collided with VER and severely damaged her bows. This forced her to return home but she remained in service for at least six months before she could be taken in hand for repairs.

The replacement of a complete bow section was necessary in the case of ACHILLES, following a collision in thick fog in the English Channel, with the supertanker OLYMPIC ALLIANCE on 12th November 1975. ACHILLES limped into Portsmouth for an assessment of the damage before heading for Devonport for repairs which were complete by the following March.

At this time, those Leanders not involved in the long refit programme had been in a short (six month) refit cycle at Gibraltar, where the completing ship was manned from the incoming frigate. This cycle was badly disrupted by commitments in the Arctic, and the more immediate need to repair damaged frigates meant that those "still in one piece" stayed in commission longer than intended.

One of the more mundane tasks undertaken by the Royal Navy, and in which the Leanders were involved, was the Beira Patrol — the British blockade of Portuguese ports in Mozambique in support of a UN Resolution to deter oil supplies reaching Rhodesia, following the declaration of UDI in November 1965. This patrol was to continue until 1975 when, to the relief of all concerned, it was withdrawn. Another patrol task was that of Belize guardship, patrolling the northernmost coast of Central America to deter an invasion of Belize by Honduran forces. There were also spells as the West Indies guardship as well as a similar role at Gibraltar. The continued commitment to NATO was honoured with the allocation of a frigate, frequently a Leander, to the Standing Naval Force Atlantic as well as the annual activation of the Naval On-Call Force Mediterranean (NAVOCFORMED). The Silver Jubilee Fleet Review at Spithead on 28th June 1977 saw a total of fifteen British Leanders on view, together with the New Zealand class member CANTERBURY and the Indian UDAYGIRI.

Following the sucessful completion of Seawolf trials with PENELOPE and the fitting of the missile system in the Type 22 frigates then building, the decision was taken to convert the "Broad-beamed" Leanders to carry both Seawolf and Exocet missiles. This combination, together with a Lynx helicopter, restored these frigates to the first-rate general purpose escorts which had been the intention of the original design in 1960. ANDROMEDA was the first ship to enter Seawolf refit at Devonport in December 1977, to be followed by CHARYBDIS in June 1979, JUPITER in January 1980 and SCYLLA the following November. HERMIONE commenced her conversion at Chatham in January 1980 and was to become the last ship to refit in the Dockyard before its closure in September 1983. The ship left Chatham on 8th July 1983 for Devonport, completing her refit there on 8th December of that year.

Prior to starting her Seawolf refit, in October 1978, SCYLLA had been used in trials involving stern refuelling from commercial tankers, in this case the BP River class ship BRITISH ESK. The tanker was slightly modified to take a hose over the stern, and by this method 100 tons of fuel were taken on by the frigate. This method was to prove highly successful in refuelling commercial ships used in support of the Falklands campaign.

The Defence White Paper of 1981 put an end to the Seawolf conversion programme when the Secretary of State for Defence announced that only the five ships currently under refit would actually be converted and the major half-life refits of the remainder would be abandoned. The conversion cost per ship had risen to £60,000,000, not far short of the cost of a new ship at that time. Consequently, JUNO, a Batch 2 ship scheduled to receive an Exocet refit, and the Batch 3 ships APOLLO, ARIADNE, DIOMEDE, BACCHANTE and ACHILLES were to remain unconverted and to head for premature disposal. Although their weapon systems were outdated there was still a considerable hull-life remaining.

HMS BACCHANTE, 17/5/82, leaving Portsmouth en route for service in the South Atlantic   Mike Lennon

Following this disastrous and ill-conceived "Review", JUNO was laid up at Chatham as a member of the Standby Squadron, to await her fate. In September 1981 the ship was towed to Rosyth where she was again laid up until work commenced to convert her to a training ship to replace the ageing Type 12 TORQUAY. By early 1984, JUNO had been gutted, her weapons and radar removed and her machinery prepared for refit. The removal of the weapons systems allowed increased accommodation and teaching space to be installed for the ship's role as the Fleet Navigation Training Ship. The refit also permitted updating of the machinery and the improvement of habitability to modern standards. While fitting-out, the ship was equipped with STWS-1 ASW torpedo mounts, to port and starboard, abaft the hangar. As a result of the removal of the Mk 10 mortar it was possible to extend the flight deck aft to permit the operation of a Lynx helicopter. However, as the hangar was converted to a

HMS JUNO, 15/4/85 showing changes effected during conversion to a training ship   L. van Ginderen

workshop, a Lynx cannot be permanently assigned to the ship. The weaponry was completed by the re-fitting of single 20 mm Oerlikon guns in the original positions on the bridge-deck; the Corvus launchers were also retained in their old positions. The electronics were updated and the old Type 978 navigation radar replaced by the modern Type 1006 set. On the forecastle a third breakwater was added, on the site of the removed 4.5 in gun turret. Following six months fitting-out, JUNO carried out engine trials in December 1984 before final docking in early 1985. Sea trials commenced on 21st February, 1985, prior to recommissioning for her new duties.

Meanwhile, BACCHANTE was declared for disposal in 1982 and sold to New Zealand, becoming HMNZS WELLINGTON. In addition ACHILLES and DIOMEDE were due to pay-off into the Standby Squadron, pending a decision as to their future. The conflict with the Argentinians over the Falkland Islands intervened and those ships slated for premature disposal became fully operational once again. Although none of the unmodernised ships were involved in the fighting they have been deployed to the South Atlantic frequently since the war ended and, because of the threat of Argentine air attack DIOMEDE, ACHILLES, APOLLO and ARIADNE were fitted with extra 20 mm guns of varying vintages. Thus, DIOMEDE was fitted with a Second World War vintage 20 mm

HMS DIOMEDE, arriving at Portsmouth 17/10/82 after service on the Falklands Islands patrol
                                                                                        L. van Ginderen

HMS ARIADNE, the last of the class to be completed, seen on 20/5/83 entering Portsmouth L. van Ginderen

gun on her stern while ARIADNE received a modern B-MARC GAM B01 gun for her stern bandstand. The fate of the four remaining unmodernised Leanders was decided in the 1984 Defence White Paper which stated that the eight frigates due for disposal under the earlier proposal (including two Rothesay class ships) would now remain in the operational Fleet.

Consequently DIOMEDE started a restorative refit at Portsmouth in September 1983, which was completed by the middle of 1984. The only external modification apparent was the "slimming" of the navigation radar platform and the update of the navigation radar. It is understood that the machinery was overhauled and the accommodation brought up to modern standards.

On 4th June 1984 ACHILLES arrived at Devonport to begin a similar refit, her first for six years. The work was completed on 12th April 1985 and was followed by a work-up period at Portland. APOLLO commenced her refit on 30th July 1984, also at Devonport, completing on 17th May 1985 after forty two weeks in dockyard hands. The refit had cost of £11,000,000. Prior to this refit APOLLO had steamed some 370,000 miles since her

HMS APOLLO, 21/5/84 leaving Portsmouth. Note single 20 mm GAM B01 gun on bandstand aft
L. van Ginderen

completion in 1972, (30-35,000 miles a year being a typical figure for a Leander), the majority of this time having been spent with the Second Frigate Squadron operating out of Portland. APOLLO was rededicated on 28th June 1985, six months after ARIADNE had started her intended forty two-week refit, which was completed in October 1985. ARIADNE had last refitted at Gibraltar in 1980 before deploying to the South Atlantic, just before the Falklands Conflict. Most of 1983 was spent as Gibraltar Guardship, followed by trips to the United States, and flood-relief work in Jamaica while West Indies Guardship. During 1987 ARIADNE was deployed to the Persian Gulf as part of Armilla Patrol Group Kilo, accompanying more than 2,000,000 tonnes of shipping through the Straits of Hormuz in a two month period.

All four vessels retained their original armament, together with the additional close-in weapons which had been fitted following the experiences in the Falklands. All except ARIADNE reappeared with a modified, "slimmed-down" navigation radar platform, although all received the Type 1006 surface search radar. ACHILLES and APOLLO both had their port side davits and motor boat removed and replaced by a light pole-derrick for the launch and recovery of the much lighter inflatable craft with which both ships were re-equipped.

With the withdrawal of the aged Wasp helicopter from RN service in March 1988 the five unmodernised Leanders were left without an aircraft, severely reducing their already limited capabilities and consequently disposals continued. ACHILLES and DIOMEDE were reportedly offered to the Greeks in 1987, but when

Pakistani frigate SHAMSHER (ex-HMS DIOMEDE), leaving Portsmouth 26/7/88   *L. van Ginderen*

DIOMEDE paid-off on 7th July 1988 it was for transfer to Pakistan, the ship commissioning as SHAMSHER on 15th July. DIOMEDE had completed a five-month tour to the South Atlantic in December 1987 where she was replaced by APOLLO; the latter returned to the UK in July 1988, undergoing a brief docking at Southampton prior to de-storing at Portsmouth. APOLLO paid off at the end of August also for sale to Pakistan, commissioning on 14th October 1988 as ZULFIQAR.

The second Pakistani Leander class frigate ZULFIQAR (ex-HMS APOLLO) leaving Portsmouth 2/12/88
    L. van Ginderen

ACHILLES continued in service with the Sixth Frigate Squadron, until the Seawolf Leanders were allocated to the squadron. The ship then became part of the Dartmouth Training Squadron, embarking midshipmen from Britannia Royal Naval College for sea-going training. In this role ACHILLES visited the West Indies and West Africa in company with BRISTOL and EURYALUS, and in May 1989 deployed to the Baltic with BRISTOL, becoming the first British warship to visit Rostock in over forty years, while the destroyer went to Leningrad. Other deployments involved attachment to the NATO Mediterranean squadron as the British representative in the 36th Activation of NAVOCFORMED. ACHILLES was paid off for disposal in January 1990, de-storing at Portsmouth before decommissioning in March, prior to being towed to Devonport to be laid up pending a decision as to her future.

HMS ACHILLES being towed from Portsmouth 27/3/90 by RMAS ROBUST to be laid up at Devonport
Walter Sartori

At the time of writing (October 1990) the futures of ARIADNE and JUNO seemed a little more settled, both having participated in the Rosyth "garage" Leander refit programme. The crew from JUNO, as the incoming ship, transferred to ARIADNE when she completed her machinery and weapons overhaul in March 1989. During the refit, which should see her remain in service until the early 1990s ARIADNE was fitted with the same type of light boat derrick on the port side as fitted to ACHILLES and APOLLO. On completion ARIADNE joined the Dartmouth Squadron, replacing the retired EURYALUS. Her first training deployment was to the Mediterranean in autumn 1989 in company with INTREPID and ACHILLES, after which cracks were found in the mortar magazine. The easiest way to empty the magazine before welding work could start was by firing off the remaining thirty-seven mortar bombs in the Portsmouth Exercise Area, an operation which took only twenty three minutes. The Training Squadron was deployed on a world "cruise" in January 1990, visiting ports in East Africa, SE Asia, the Aleutian Islands, the Pacific coast of Canada/United States, Mexico, the Caribbean and the Azores before returning home in July.

JUNO, while Fleet Navigation Training Ship, undertook trials with the Type 2050 hull-mounted active sonar due for installation in the Type 23 frigates, completing the work in late 1986 six months ahead of schedule. After this the ship suffered several minor accidents, including grounding in the Solent during December 1986, which required docking at Portsmouth for replacement screws to be fitted; nearly a year later the Type 21 frigate ACTIVE collided with JUNO off Ailsa Craig, fortunately resulting in only minor damage. In early 1987 JUNO

HMS JUNO, 8/6/87 at Montreal. Her armament has been reduced to two single 20 mm guns abreast the foremast and two triple STWS-1 torpedo tubes abaft the hangar   L. van Ginderen

participated in the Dartmouth North American cruise in company with FIFE, the pair being joined by EURYALUS and APOLLO in the later stages of the deployment. After periods at Portland in 1988 JUNO prepared for her refit at Rosyth which was expected to last from March until late 1989. The refit involved the removal of the ASW torpedo tubes, the repair/replacement of various pumps and other machinery, as well as ten other additions and alterations. Extensive work was carried out on the galley deck which was completely refurbished, while the engine room was also overhauled. An unexpected problem arose with the hydro washing of the ship's hull, as it was necessary to ensure that the washings did not reach the Firth of Forth due to the toxic nature of the erodible, anti-fouling paint used; disposal of the waste had to be very carefully controlled. On completion of the refit JUNO was manned by the crew of SCYLLA for sea trials and service, while the Seawolf Leander started her own refit at Rosyth. Having completed the refit and rejoined the Sixth Frigate Squadron it would appear likely that JUNO will remain in service for several more years.

CHAPTER THREE

BATCH 1 IKARA CONVERSIONS

The Leander design had been completed just before the large scale introduction of guided missiles, the first seven of the class being completed without the Seacat missile system, despite the fact that the system was already at sea in four later Battle class AD conversions and the new destroyer DEVONSHIRE. Consequently, by the time that the first group of Leanders were approaching ten years of age their weapons and electronics systems were in need of radical updating to avoid block obsolescence. In 1968 therefore, it was decided to convert twelve ships to carry the Ikara AS missile system. Ikara was an Australian quick-reaction missile designed to deliver AS homing torpedoes at a target detected at some distance from the mother ship. Under development in Australia since the late 1950s, the Royal Navy ordered a trial batch of missiles in February 1963, for evaluation. After the trials it was hoped that it would be possible to develop the design further in the UK and, with Australian Government agreement, eventually build it under licence in Britain. The US Government contributed £1,428,000 towards the development costs.

By 1966 work was underway to develop launch gear for Ikara to be fitted in British frigates, and two years later the early Leanders were selected for conversion; later, the number to be converted was reduced to eight. Ikara had been at sea for some time in the Australian Navy, fitted on the quarterdeck of their River class frigates, but to do this with the Leanders would have caused operating problems with the Mk 10 mortar, as well as interfering with helicopter handling.

To obviate any delays in the refit programme, the American Stanwick Corporation were given a contract in October 1969 to study the Royal Navy refit schedule and make recommendations. Their task was to identify potential hold-ups and to advise on the size and make up of the workforce.

Modifications had been made to Leander class ships as they were building, following operational experience with those ships of the class already in service. Consequently, LEANDER was probably the least altered ship in the class when she entered refit and still shipped the "temporary" single 40 mm guns aft instead of the Seacat system. With the Seacat ships, there were variations in the type of Seacat director and its position and the opportunity was taken to update and standardise this and other equipment.

The Ikara conversion involved the removal of the 4.5 in gun turret, its training gear, shell handling room, magazine and the associated MRS3 director on the bridge. Modifications were made to the foremast, the Type 965 aerial was removed from the mainmast and the 40 mm guns from the hangar roof. Externally the hull was little altered, though on some later vessels the bow structure and bilge keels were modified. The so-called LONDONDERRY-bow was fitted to all ships to improve the flow of water over the sonar dome, and the bilge keels aft were removed to reduce hull-vibration which occurred in heavy weather with the use of the fin-stabilisers.

Starting forward, the foremost breakwater was retained but, from the forecastle break aft to the bridge, the Ikara launcher and handling room were constructed. The launcher itself was housed in a circular deckhouse or "zareba", and the handling room was built between the zareba and the main superstructure.

Designated GWS40, Ikara had been first fitted in BRISTOL; however the system to be fitted in the Leander conversions was redesignated GWS41. The building of BRISTOL was a somewhat protracted affair and the conversion of LEANDER was completed before trials of Ikara could begin in the former ship. As a result, it was LEANDER which undertook not only "First of Class" trials following the refit, but also "Ikara First Fit" trials.

Ikara consists of an aerodynamic missile within which a Mk 44 or Mk 46 torpedo is carried. On launching the solid propellant motor boosts the missile away from the ship, from where it is guided towards the target by the command guidance system. Upon reaching the vicinity of the target the homing torpedo is released, parachuting to the surface and on entering the water the torpedo commences the search mode, either passively or actively, to attack the target. The Mk 46 has multiple re-attack capability, as well as the choice of active or passive search, while the older Mk 44 relies purely on active homing. Ikara has a range of about 14 nautical miles which is the approximate range of the ships' sonar. However missile guidance can be controlled either by the firing ship or by others within the task force thereby increasing, in theory at least, the range of the weapon. In practice, however, range is restricted to that of the carrier-missile. The torpedo is already fitted into the missile in the magazine but, to permit easy handling before reaching the launcher, the wings and tail are fitted only just prior to launching. When not in use the "zareba" is protected by a hinged weatherproof cover.

As the handling room was built on to the front of the bridge superstructure it was necessary to move the forward RAS derricks. These were replaced by a hinged jackstay just to port of the centreline, on the roof of the missile handling room. The oil fuel filling pipes were stowed athwartship aft of the breakwater. On the port side of the deckhouse a platform was built over the deck to the port side of the ship, for use when taking stores from the forward RAS point.

The weapons fit of the Ikara ships included two Bofors 40 mm AA guns, one to port and one to starboard, just forward of the foremast. These were mounted on sponsons built on the forward superstructure at bridge level, above the liferafts, giving the ships a low-level operations capability. Ready-use ammunition lockers were fitted inboard of the sponsons.

With the 40 mm guns removed from the hangar roof, it was possible to alter its layout to accommodate two quadruple Seacat launchers — one each side, to port and starboard. The GWS 22B director for the system was fitted aft of the mainmast, on the centreline. In those Leanders which had carried Seacat when completed the director had been fitted on the starboard side of the hangar roof and as ships converted the director was repositioned on the centreline, thereby restricting its ability to deal with targets forward of the ship. Consequently, the Bofors 40 mm guns were expected to deal with forward air targets. The Wasp helicopter and Mk 10 mortar were retained, although the latter was updated, and consequently the flight deck and helicopter handling arrangements remained unchanged.

HMS LEANDER, 18/7/79 as an Ikara armed, specialist ASW frigate. Note "headlamp" ESM at foremast head
L. van Ginderen

The logic of fitting Ikara in lieu of, rather than in addition to, the 4.5 in gun turret has been questioned. It could be argued that to fit Ikara on the quarterdeck, as in the Australian frigates, would have meant the loss of the close range AS capability of the Mk 10 mortar and because of space limitations helicopter operations would have been impaired. Alternatively, to fit Ikara amidships as in the Australian Charles F Adams class destroyers would have required radical alteration of the superstructure. As it was intended that Ikara-converted Leanders would operate within a task group, there were also doubts about the need to retain helicopter facilities, at the expense of their general purpose capabilities. The helicopter could, of course, deliver AS torpedoes at contacts made by other ships in a task group. A further advantage was that the forward Ikara magazine capacity probably enabled the British ships to carry more reloads than their Australian counterparts.

At the stern, VDS was retained in those ships in which it had been fitted originally, and added to those not so equipped, taking sets from Leanders of Batches 2 and 3; the Type 182 towed decoy was also retained at the stern.

The electronics fit for Ikara consisted of a small aerial on the forecastle just aft of the capstans, a radome on the bridge top where the MRS3 director had been fitted and a second aerial, mounted on a short pole mast between the radome and the foremast. These formed the command guidance system of the missile. The foremast was slightly remodelled to take platforms for the aerials for the SCOT (Satellite Communications Onboard Terminal) which improved UK naval communications throughout the world. Fitted halfway up the mast, angled forward on the starboard side and aft on the port side, each platform carried a raised circular base on which was to be fitted the actual SCOT radome. The radome was not fitted all the time, but principally when the ship was on overseas deployment or involved in exercises.

HMS AURORA at Chatham 16/3/76. An Ikara missile can be seen on the launcher in the zareba fitted forward of the bridge. Missile guidance aerials are fitted under the randomes on top of the bridge and on the short pole mast immediately forward of the foremast   R. H. Osborne collection

The navigation radar platform was angled out to the side and the Type 974 radar replaced by Type 978, and eventually by Type 1006, when the latter set became available. Furthermore, as the Type 994 came into production it was retrofitted, though retaining the original Type 993 aerial at the top of the foremast. There was a difference at the foremast top, however, where the so-called "headlamp" aerial was fitted. This stabilised and trainable aerial was to form a part of the ESM suite, possibly to detect emissions from submarine attack periscope radars.

Moving aft to the mainmast, the AKE-1 aerial for the Type 965 radar was removed and replaced by the Cossor Mk 10 IFF antenna Type 1010. On either side of the mast, at shelter deck level, sponsons were built out to the ship's side to carry the Corvus rocket launchers with which all of the class were to be fitted. Some ships, like ARETHUSA, had carried Corvus for some time prior to conversion. The ships' sonar suite comprised the hull-mounted Types 184, 170 and 162 sonars, which were retained in addition to the VDS. The positioning of the ships' boats and liferafts remained the same as in the first and second batches but in addition an inflatable was carried on the hangar roof, forward of the starboard Seacat launcher. The securing points for the RAS jackstay and its supporting cables were retained on the flight deck. Few structural alterations were made to the main superstructure but forward of the mainmast a small flying bridge was fitted either side of the emergency compass platform, at the extremes of which were fitted whip aerials. In all, the changes, plus ballast, resulted in an increase in displacement from 2,380 tons standard, 2,860 tons full load, to about 2,500 and 3,000 tons respectively, but with increased automation it was possible to reduce the complement to about 250.

Internally the steam plant was improved with the installation of a steam-atomisation combustion system; the boilers were modified to burn Dieso only, permitting the fuel tanks to be water-displaced, and the air-conditioning was improved. The internal arrangement of the Ikara conversions was altered further by the enlargement and reorganisation of the operations room. At the same time living standards were improved to the standards then required. The fitting of the additional electronics and the improvement of habitability meant that an increase in generating capacity was required and consequently, the two original 300 kW diesel generators were replaced by two units producing 500 kW each. With the removal of the Type 965 radar the space occupied by the associated electronics became available for the Ferranti FM1600 computer which was the basis of the ADAWS 5 (Action Data Automation Weapon System) which had evolved from the ADA (Action Data Automation) system fitted aboard the carrier EAGLE.

ADAWS 5 permitted a multi-ship tactical command and control for the Ikara missile from launch to the dropping of the torpedo in the vicinity of the submarine target. Using information from the ship's own sensors (sonar, radar and HF/DF etc.) the computer could automatically activate the missile on the launcher, and then train it to the correct bearing and elevation to reach the calculated target zone. Instructions were passed by the command guidance link on the bridge to the missile and once in flight, course corrections were made using the radio link. Additional information, and indeed guidance, could be provided by other ships in a task force. When the computer "decided" that the missile had reached the dropping zone, the torpedo was released and parachuted into the water to commence its attack. The body of the missile meanwhile flew clear of the area. The advantage of Ikara over the MATCH concept was the all-weather capability of the former, despite the greater range and flexibility of the helicopter.

Conversion of the eight Leander class ships was originally costed at £4,900,000 each, with a limit of £5,000,000 per ship. Serious planning of the refits had commenced in January 1969 when a "Project Manager Ikara Leanders" was appointed. It was intended that each ship should spend about two years in dockyard hands, which would include a nine month docking period. LEANDER, AJAX and DIDO had all been laid down as Rothesay class ships and, although modified while building, still retained some of the compartmentation of the original design. These structural difficulties led to problems in the re-fitting of the water-compensated fuel system and increased docking time was required for the three ships. Following the docking and installation of Ikara (GWS41), ADAWS 5 and the improved generating and air-conditioning, a further nine months was required to "Set to Work" the first two systems, as well as completing the outfitting.

The refit programme commenced with LEANDER in 1970 and was completed in September 1978, when DIDO recommissioned. It was intended that the Ikara conversions would be divided into two sub-batches, 1A and 1B, the latter incorporating alterations and improvements learned from the Batch 1A refits. LEANDER, AJAX and GALATEA were selected as 1As for refit at Devonport, while the remainder became Batch 1B. The majority of the refits were undertaken at Devonport Dockyard although, because of the work-load at that yard, ARETHUSA was converted at Portsmouth and AURORA at Chatham. It had been originally intended that DIDO should be refitted at Chatham, but she was reallocated to Devonport.

HMS AURORA off Portland in 1976. The SCOT antennae can be seen fitted to port and starboard, halfway up the foremast
R. H. Osborne collection

HMS GALATEA, 9/74 on completion of her Ikara conversion　　　　　　　　　　　　R. H. Osborne collection

LEANDER satisfactorily completed her Dockyard Sea Trials, including Ikara launcher blast trials, in June 1972, and following work-up recommissioned in December the same year after one hundred and thirty six weeks in dockyard hands. As previously mentioned LEANDER recommissioned after her major refit six months before the last of the class, ARIADNE, was completed. AJAX, the next to be taken in hand in 1970, completed in February 1974 while GALATEA completed the following September, followed by NAIAD in June 1975. EURYALUS was next in line at Devonport while AURORA started at Chatham and both recommissioned in March 1976. ARETHUSA paid off for refit at Portsmouth in 1973 and recommissioned in April 1977, having left drydock in June the previous year. The last to be converted was DIDO which had been in service for fifteen years before paying off, the last 40 mm-gunned, non-Seacat-armed Leander to be modernised.

HMS LEANDER leaving Portsmouth in 6/76. Note bow damage caused during the "Third Cod War"
　　D. A. Sowdon

HMS LEANDER, 18/12/84. The "headlamp" has been replaced by a new ESM array at the top of the foremast
L. van Ginderen

Post conversion there were small alterations to various ships, for example several had a small, additional deckhouse fitted just forward of the foremast. All ships, except DIDO, "lost" their VDS transponders — AJAX was probably the first and was seen in August 1983 without the "fish" or cable for the VDS, but still retaining the pantograph. The so-called "headlamp" ESM aerial fitted to the Ikara conversions (also seen on the "Broadbeam" ANDROMEDA in early 1976) had been removed by August 1983 from all but NAIAD, which lost it during her final refit. A shorter, lighter, aerial was fitted in lieu of the "headlamp" in all ships. An unseen improvement to the ECM battery commenced in 1978 with the replacement of the Knebworth chaff rockets by the BBC — Broad Band Chaff — rocket. Fired from the same Corvus launcher, one BBC rocket was the equivalent of three Knebworth rounds, with a payload of 6 kg. The improvement was necessary to combat the more advanced radar and infra-red homing systems being fitted in modern anti-ship missiles.

Now that they were specialised ASW ships with only Seacat and a pair of 40 mm guns as self-defence, these Ikara-armed ships needed to operate in mixed squadrons to provide some defence against air and surface attack. Consequently, they were allocated to squadrons containing a mixture of GP frigates. A typical example was the Third Frigate Squadron which in September 1974 consisted of the Ikara-armed LEANDER (leader), and the unconverted DIOMEDE (half-leader) and ACHILLES, both "Broad-beam" ships, together with the Rothesay class ships LOWESTOFT and FALMOUTH. This group, in company with the SSN WARSPITE and led by the cruiser BLAKE, took part in a First Flotilla deployment to the Far East in 1974/75; one of a series of such cruises which had commenced in 1973, following the withdrawal of permanent British squadrons overseas. The 1981 White Paper was to stop future deployments on the grounds of cost, despite their role, albeit secondary, as successful defence sales media. However, the deployments were recommenced in 1983 when a group, including AURORA, left that September for a seven month visit to Australia and South East Asia. Another change effected was the abandonment of the mixed squadron system and its replacement by squadrons made up of ships of one type. As a consequence, the Ikara Leanders were allocated to the First Frigate Squadron while Exocet and Seawolf ships were transferred to the Seventh and Eighth Frigate Squadrons respectively.

Under the cost-cutting Defence Review of 1981, several of the Leanders were listed for premature disposal, including the Ikara-armed DIDO. The last to complete this conversion, DIDO paid off from Royal Navy service

HMS DIDO, 7/5/79
Mike Lennon

HMS NAIAD 13/5/83 at Walsoorden. The two Seacat launchers, to port and starboard, and the centreline GWS 22 director can be seen on the hangar roof
L. van Ginderen

in June 1983 and was transferred to the Royal New Zealand Navy on 18th July that year becoming HMNZS SOUTHLAND. NAIAD was due to be transferred to the Standby Squadron in 1983 but, because of the increased commitments following the Falklands war, this order was rescinded and the ship remained in service. Subsequently, in September 1983, NAIAD was taken in hand at Devonport for a £10,400,00 refit, (coincidentally, the cost of her conversion to June 1975) her first in five years, thus becoming the first Ikara Leander to undergo her final refit. This was completed in early June 1984 after thirty nine weeks in dockyard hands, a week and £1,600,000 less than the original estimated time and cost (£12,000,000). A period of work-up at Portland followed the refit.

AJAX was not so lucky, because in late 1984 it was announced that she would pay-off for disposal in 1985. In the summer and autumn of 1984 AJAX became the first Ikara Leander to deploy to the South Atlantic. During the patrol she underwent a short assisted maintenance period alongside the RFA repair ship DILIGENCE. AJAX could not have had a more auspicious end to her Royal Navy career than when she acted as escort to the Royal Yacht BRITANNIA on the tour of Italy undertaken by TRH's the Prince and Princess of Wales in April/May 1985. Following a final ASW exercise in the North Atlantic AJAX paid off at Devonport on 31st May 1985, and decommissioned on 10th June for disposal, to await that last trip to the shipbreakers' yard. However, while awaiting a final decision as to her fate, the ship was inspected by representatives of the Thai Navy with a view to purchase but, unfortunately, the Thais decided that the ship was not in good enough condition and the possible sale fell through. A further reprieve was announced during June 1985 when AJAX was allocated to the role of static training ship at Devonport to replace the older Type 61 frigate SALISBURY, which was subsequently sunk as a target on 30th September that year. AJAX was then used for the training of seamen in the handling of RAS gear and other deck equipment.

AJAX continued in this role until her replacement by a converted former dockyard fuel lighter in 1987 which took the frigate's name. The old ship was then laid up at Devonport until February 1988 when towed to Millom in Cumbria for breaking. Although she arrived off the Duddon only a few days later, the frigate was unable to cross the bar and was towed to Barrow for lightening and to await a suitable tide; the ship finally arrived at Millom in late July 1988.

HMS AJAX leaving Devonport on 15/2/88 en route for Cumbrian shipbreakers at Millom   *B. Sullivan*

    The remainder of the Ikara conversions followed NAIAD into a final refit. AURORA commenced a docking period at Devonport in September 1984 which was completed on 28th June 1985, and followed by her rededication on 5th September. During this £7,500,000 refit, the decision was taken to add a further 100 tons of ballast to the ship, thereby extending her time in dockyard hands. LEANDER was similarly equipped during an extended docking period which commenced at Devonport in January 1985 and was completed during the summer of that year.

    As part of a cost-efficiency comparison exercise EURYALUS and ARETHUSA were selected to undergo "comparator" refits, the former in a civilian yard, the latter at Devonport. The commercial refit was put out to tender and won by Tyne Shiprepairers at South Shields, who took EURYALUS in hand in November 1984. Prior to starting this refit, the frigate's VDS gear was removed and she was towed from Devonport to Portland and then on to the Tyne by the RMAS tug ROBUST. The refit which started on 25th November was supposed to have been completed on 10th September 1985, but was extended until 10th October to allow the installation of a further 100 tons of ballast. However, further unexpected delays conspired to keep EURYALUS under refit on the Tyne until the end of November 1985 when she sailed for Devonport. As soon as the frigate arrived at Devonport she entered the dockyard for a two-week assisted maintenance period so that additional work could be carried out.

The refit of ARETHUSA started on 29th May 1985 and was to include work not undertaken during the refit of EURYALUS. ARETHUSA was equipped with the Type 2031 'I' towed array sonar set removed from the Rothesay class frigate LOWESTOFT which had paid off for disposal in March 1985. To accommodate the towed array, the former VDS well was plated-in completely and the cable-drum and its associated winch gear fitted right aft, on the centreline. The Mk 10 AS mortar was removed and its well plated over to provide a sonar operations room for the new equipment. As a result, when ARETHUSA re-entered service in March 1986 on completion of the refit, she became the only Batch 1 Ikara-armed Leander to be equipped with towed array sonar.

HMS ARETHUSA leaving Antwerp 7/9/88. She was the only Ikara-armed Leander to be equipped with Type 2031 'I' towed array sonar. The winch and cable drum can be seen right aft on the quarterdeck. To accommodate this sonar the Mk 10 AS mortar was removed and the mortar and VDS well plated-over   L. van Ginderen

By July 1984 the VDS fitted to GALATEA had been removed and, as in AURORA, an additional close-in weapon, a 20 mm Oerlikon gun, fitted on the starboard quarter. Prior to beginning an eight-month attachment to STANAVFORLANT, in September 1984, GALATEA had had the distinction of being the last Royal Navy ship to undergo an extended docking period in Gibraltar Dockyard before the yard was transferred to commercial interests. On completion of the NATO deployment, in mid 1985, GALATEA arrived at Devonport for refit, before starting her last year in commission.

HMS LEANDER arrives at Portsmouth 1/8/86 to pay off into the Standby Squadron   L. van Ginderen

The paying-off of Ikara conversions continued with LEANDER (31st July 1986) and GALATEA (August 1986), both being laid up at Portsmouth in the re-formed Stand-By Squadron, officially at thirty days' notice for sea. Portsmouth had at this time been designated the port at which vessels should be prepared for disposal. On 1st April 1987 however, both ships were reduced still further and GALATEA was prepared as a target and then towed from Portsmouth on 13th June 1988 by RMAS ROLLICKER, to be expended during Exercise JMC 882 in the North Sea later that month. Prior to departure GALATEA's name and pennant number were erased as part of a new policy which dictated that target ships should be sunk anonymously. LEANDER, also now unnamed, was towed into Portsmouth harbour from the dockyard for lay-up prior to disposal, also as a target, during exercises in September 1989.

HMS LEANDER being towed from Portsmouth in early 9/89 by RMAS ROLLICKER and RMAS POWERFUL en route for NATO exercise "Sharp Spear" during which the frigate was hit by one Sea Dart and three Exocet missiles before being sunk by a 1,000 lb bomb   Gary Davies — MG Photographic

AURORA paid-off on 18th April 1987, one year earlier than intended, due to the manning crisis within the Royal Navy at the time, and after de-storing in Portsmouth was laid-up in Fareham Creek for disposal. AURORA was sold to Francisco Mala SA of El Ferrol who then resold the vessel to Devonport Management Limited, the consortium which took over the dockyard in 1987. DML brought the ship as a speculative venture with several potential customers in mind and were prepared to refurbish the ship to the requirements of the new owners. Following purchase by DML, AURORA was towed from Portsmouth on 1st August 1988; however no buyer was found and she was sold for scrapping. Subsequently she arrived at Barrow on 18th July 1990 for lightening before being towed to Duddon Valley Shipbreakers at Millom on 19th August. After grounding, she was eventually berthed on 6th September.

HMS AURORA aground off the old jetty at Millom 27/8/90   David Crowdy

NAIAD paid-off at Portsmouth a fortnight after AURORA, and following a short period in reserve was selected as a floating test-bed for trials of design and construction developments resulting from the lessons of the Falklands operations. The three part tests were to involve fire, shock and blast trials, the first at Portsmouth between July and October 1988 after earlier mining trials. The ship lost her name and number before the trials commenced, being referred to instead as "HUL-VUL" for the duration of the tests. Part two was conducted at ARE Dunfermline in the spring of 1989, the hulk having been towed there from Portsmouth by RMAS ROBUST

HUL VUL (ex-HMS NAIAD) arriving at Portsmouth 5/8/89 after trials in Scottish waters   Walter Sartori

in March. HUL-VUL arrived back at Portsmouth on 5th August 1989 in tow of the RMAS tugs FOXHOUND, BUSTLER and ROLLICKER for the final blast and fragmentation trials scheduled for autumn 1989. Several concepts were being considered during the trials which involved the reconstruction of part of the ship to resemble a Type 23 frigate. These included the evaluation of the Wormauld "Status" damage control monitoring system, a blast-proof control room, fire-resistant computers and a shock-absorbing deck. In addition Sea Gnat and Super RBOC chaff rockets were tested against a "live" Exocet missile, off the coast of Scotland during the summer of 1989. Lessons learnt during the trials will presumably be incorporated into new building and possibly lead to retrofitting during refits of existing ships. On completion of these trials the hulk was due to be sunk as a target in the Western Approaches during 1989. However, HUL VUL was retained and fitted with a large box girder bulge along each side. The function of this trial is to test the viability of incorporating similar structures into the passageways of new ships thereby increasing longitudinal strength without restricting movement around the ship. Trials continued until the Autumn of 1990, when, as planned, she was sunk as a target.

HMS EURYALUS. In the Ikara conversion a single Type 1010 IFF aerial replaced the Type 965 AKE-1 aerial (with IFF) fitted at the top of the mainmast since completion   R. H. Osborne collection

The final pair of Ikara ships, EURYALUS and ARETHUSA, paid-off at Portsmouth at the beginning of April 1989, only three years after completing refits which had been expected to extend their lives for five years. Following her last refit EURYALUS was deployed to the South Atlantic in July 1986, returning to the UK on 18 December that year. Attachment to the Dartmouth Squadron in 1987 followed, involving duty as training ship for midshipmen and included cruises to the Baltic and West Indies. EURYALUS retired following her return from the Dartmouth West African cruise in February/March 1989, having steamed almost 700,000 miles during her twenty-five year service life, more than any other Leander. Following de-equipping and de-storing EURYALUS was sold to Devonport Management Ltd and laid-up in Fareham Creek. As no buyer could be found, she was resold to Duddon Valley Shipbreakers, and arrived in tow at Barrow on 26th September 1990 to be lightened before proceeding to Millom.

HMS EURYALUS lying at Barrow 14/10/90   *Michael Crowdy*

ARETHUSA, whose £12,500,000 refit had restored some operational capability, participated in several exercises during her last three years, including Caribtrain '87 with AURORA and ARGONAUT in January to April 1987, JMC 882 in which GALATEA was sunk, as well as training sessions at Portland. ARETHUSA underwent a final docking period at Devonport, completing an eleven-week DED on 13th November 1987, during which a 5-ton boat crane was fitted on the port side and a new type of flexible coupling fitted to her starboard main engine. ARETHUSA was still in Portsmouth Dockyard in April 1990, probably to end her days as a target. The withdrawal of these two ships saw the end of the deployment of the Ikara missile in Royal Navy service.

Although very adequate ASW ships when first converted, the Ikara Leanders were fitted with an AS mortar which first entered service about 1951 in RELENTLESS and was obsolete by the time the refit programme commenced. Furthermore, seven men were required to operate the mortars, compared with the crew of three used in the Australian version of this weapon. The VDS, which was never completely satisfactory in service, is understood to have proven unsuitable for operations in the Eastern Atlantic, where the Royal Navy would expect to conduct most of its operations in the event of war. Hence the sonar was removed from these ships before or during their final refits; indeed, VDS was removed from AJAX as early as mid 1983. Subsequently, this ship also had her VDS well plated-in at the transom, probably prior to service in the South Atlantic, while the remaining ships stayed in service with an empty well. The Wasp helicopter, while still useful, was a rather elderly airframe and considerably inferior to the Lynx carried by the Exocet and Seawolf conversions. However, to carry a Lynx would have required the removal of the Mk 10 mortar and considerable alteration to the hangar, modifications which were not made due to age and cost. Finally, their main armament, the Ikara missile, fell out of favour in the late 1970s because its range was less than that at which modern sonars could detect relatively noisy nuclear-powered submarines and also because it could not carry the new British Stingray lightweight torpedo. Consequently, the AS functions of the Ikara Leanders were gradually taken over by the larger and more capable Type 22 frigates, able to carry two Lynx helicopters, which have a far greater range and payload than the elderly Wasp.

CHAPTER FOUR

BATCH 2 EXOCET CONVERSIONS

It was apparent at the time that LEANDER was taken in hand for conversion that it was necessary to update the rest of the class to prevent wholesale obsolescence of the greater part of the Royal Navy's surface escort fleet. The effectiveness of suface-to-surface missiles had been brought home by the sinking of the Israeli destroyer EILAT and, in a "Supplementary Statement on Defence Policy" on 28th October 1970 it was announced that the strike capability of surface ships would be increased to compensate for the reduction in the carrier force made by the previous Government. Negotiations were already underway with the French for the purchase of 300 Exocet missiles, their container/launchers and the fire-control equipment. Thus, in March 1973 the later County class destroyer NORFOLK was refitted with a quadruple Exocet launcher in place of "B" gun turret. The ship carried out firing trials with the missile on the French ranges, off Toulon. Evidence of the destructive capability of the Exocet system was seen later, with the use of the old Type 15 frigate UNDAUNTED as a target ship and, later still, during the Falklands campaign. As with Ikara conversions, The Batch 2 vessels were sub-divided into two sub-batches. Batch 2A consisted of CLEOPATRA, PHOEBE, SIRIUS and MINERVA, and Batch 2B ARGONAUT, DANAE and PENELOPE. Originally, JUNO was to have been converted at Devonport to a Batch 2B, but this was cancelled and the frigate was later converted at Rosyth to a Training Ship.

The first ship to be converted was CLEOPATRA, commencing at Devonport in late 1973. The planning of the refit, which was to set the programme for the remaining Exocet conversions, began in 1971. The following year the "Project Manager" was appointed and at the same time the Statement of Requirements from MoD(N) was converted into an initial, but detailed, work programme by the Design Division in Devonport Dockyard.

HMS CLEOPATRA, 1/76 undergoing a short work-up at Portland prior to her first deployment following a two-year long Exocet conversion
MOD(N)

The conversion was more extensive than the Ikara refit, involving the removal of all the ship's existing weapon systems and some of the electronics. Forward, the 4.5 in gun mounting, and the associated MRS3 director on the bridge top, were removed. The shell-handling room and magazine were converted to take Seacat missiles and the shell-hoist modified to transfer missiles to the upper deck. The forward part of the old magazine was converted into a non-water compensated Dieso tank. The superstructure forward of the mainmast was altered while the Seacat launcher, together with its director on the hangar roof, and the Mk 10 mortar were removed. Where VDS gear had been carried, this was removed in the seven ships which were actually refitted to this standard and the well plated-in.

On the forecastle the forward breakwater was retained and an additional curved breakwater fitted in front of it. This was designed to deflect breaking waves away from the forward Seacat and Exocet launchers fitted in front of the bridge. Aft of the breakwater, on the forecastle, a platform was constructed on which was mounted a quadruple Seacat launcher, in front of, and between the two twin Exocet launching ramps. Behind each pair of ramps a blast deflector was fitted to direct missile efflux away from the ship; between the blast deflectors was a whip aerial which formed part of the missile command link.

Unlike Ikara, Exocet does not require any link with the firing-ship after launching — that is, it is a "fire and forget" missile. The main requirement is for the firing-ship to be facing within 30° of the target, as the missile has only a limited ability to turn in flight. Target information acquired by the ship's sensors is relayed to the missile, activating the internal guidance system. Each missile is housed in its own container/launcher made of light alloy, 3 mm thick, but ribbed for strength. The container is filled with an inert gas at low pressure to provide a dry, stable environment.

Having acquired the target information and warmed up the gyros prior to launch, the missile is fired clear of the ship at a speed just under Mach 1 by a solid-fuel booster, after which a sustainer rocket takes over. The missile reaches a height of 30 m before dropping to 15 m when about 4 km from the launching-ship. At a distance of 24 km from the ship the missile descends to approximately 7 m, though this depends on the sea-state. At this stage the active radar in the nose of the missile becomes activated and the search for the target commences. This may entail small course corrections before the missile hits the target. The warhead is semi-armour piercing and contains about 65 kg of high explosive, with both contact and proximity fuses.

The total weight of the missile is 1,750 kg — four missiles therefore weighing 3.5 tonnes and, even allowing for the steel mounting ramps and equipment at 15 tonnes, the system weighs considerably less than the 45

HMS DANAE arriving at Antwerp 12/2/88. The forward Seacat launcher and two twin Exocet missile container/launchers can be seen clearly ahead of the bridge. Two single 40 mm guns have been fitted abreast the foremast
L. van Ginderen

tonnes of the 4.5 in gun turret alone, excluding ammunition. However, all this weight is above deck level, without the compensating weight of the shell-handling room and magazine below. This fact later led to modifications in the Batch 2B conversions, where topweight became a critical factor. Although the anti-ship capability of Exocet is considerably greater than that of the 4.5 in gun, the former system suffers in that no reloads are carried and, until recently, the launcher/containers had to be placed on board by crane while alongside in port.

The bridge superstructure remained unaltered so it was possible to retain the RAS derricks in their original positions. On the bridge top the MRS3 director was replaced by the GWS 22D director, mounted on a low platform surrounded by a guardrail, for the forward Seacat launcher. Just aft, on either side of the foremast, sponsons were built for the single 40 mm guns that supplanted the 20 mm Oerlikons carried originally.

The ships' boats and liferafts were retained in their original positions on the forward part of the superstructure, which was considerably altered aft of the funnel. Abaft the boiler room uptakes flying bridges were constructed, to port and starboard, to carry SCOT radomes and a small deckhouse was fitted on the front of the mainmast, between the flying bridges. Corvus rocket launchers were carried on extensions of the shelter deck, abreast the mainmast. In the first two conversions, CLEOPATRA and PHOEBE, the shelter deck was built out to the ships' sides and carried aft along the length of the hangar. Two sets of triple tubes for the STWS-1 ASW torpedo system were mounted on this shelter deck alongside the hangar. The positioning of these mounts may have accentuated the stability problem because in the rest of the conversions the STWS-1 tubes were mounted on the main deck, to port and starboard of the after end of the hangar. In the later ships the Corvus launchers were carried on sponsons abeam the mainmast, in the same position as the Ikara conversions. In subsequent refits, the STWS-1 tubes of CLEOPATRA and PHOEBE were lowered to the main deck.

HMS DANAE at Portsmouth 4/5/84. Note plated-in stern, single 20 mm Oerlikon gun on the starboard side of the quarterdeck, STWS-1 torpedo tubes abaft the hangar and a Lynx helicopter on the flight deck
L. van Ginderen

STWS-1 (Shipboard Torpedo Weapon System) Mk 1 is a British version of the US Mk 32 triple torpedo mount, fitted instead of the obsolete Mk 10 AS mortar. Developed by Plessey, STWS-1 differs in several ways from the American system, principally in that remote selection, pre-setting and firing is possible. An interface between the Action Operations Room, sonar and the mount gives the system a quick reaction capability for short range AS defence, with an increased probability of target destruction. CLEOPATRA was the first British ship to be fitted with the system. The mount consists of a stationary base bolted to the deck, on top of which a trainable, cast aluminium pedestal, capable or rotating through 190°, is fitted. Surmounting this are the three torpedo tubes, normally kept loaded for fast reaction. The tubes are heated to protect the torpedoes from the elements, and firing involves the use of high pressure air. The mounting is relatively light and there is no requirement for deck-strengthening.

The hangar of the ''Exocet Leanders'' was enlarged to take the Lynx helicopter, thereby allowing the mounting of two Seacat launchers on the hangar roof, one on either side at the after end. Aft of the mainmast was built a small deckhouse, and this was surmounted by the GWS 22C director. In this deckhouse were contained independent controls for the after Seacat mounts. Each Seacat launcher had a computer and radio link to the director allowing automatic control. Furthermore, the updated Seacat system allowed the launchers to operate independently to cope with threats from both beams at the same time. It also enabled the Seacat to remain operational in the event that the centreline director was disabled, and so give a greater flexibility to the system compared to that in the ''Ikara Leanders''.

The removal of the Mk 10 mortar and its well greatly increased the flight deck area which was essential for handling the larger Lynx helicopter. With the mortar well and, where fitted, the VDS well plated-in, extra accommodation was available at the stern below decks. Aft of the flight deck, the ships retained the towed decoys on the port side while on the starboard side a Gemini inflatable was stowed.

The Lynx which has a length of 15.9 m and a rotor diameter of 12.74 m, compared with 9.2 m and 9.78 m respectively for the Wasp, has a greater payload capacity, is faster and has a greater range than the Wasp. In addition the Lynx carries radar, thereby giving the aircraft an all-weather capability. The increase in size over the Wasp necessitated the use of an improved deck securing system which led to the development of the Harpoon system. This is a retractable hydraulic claw which engages in a honycomb grid, fitted into the flight deck, in the centre of the landing markings. This mechanism permits the aircraft to be secured on deck for up to eight hours without engines running.

Various changes were made to the electronics so that in place of Types 974 or 978 navigation radar, the new Type 1006 set was installed. The Type 965 radar and its associated AKE-1 aerial was retained, and communications improved with the fitting of SCOT, although as with the Ikara ships the terminals do not seem to be carried permanently. The foremast top was also altered to take larger and heavier ESM arrays. Several of the conversions, including CLEOPATRA, PHOEBE and MINERVA, carried a "cone and arrow" array for passive direction finding, and interception of enemy radar emissions. The sonar suite was improved by the replacement of the older Type 184 with the new solid-state Type 184M, and the entire Action Information system improved with the introduction of CAAIS — Computer Assisted Action Information System — based on the Ferranti FM1600B computer. The result of all these improvements, together with enhanced domestic arrangements, meant that it was necessary to increase the generating capacity to 2,000 kW. The changes also meant that the ships' complement could be reduced by forty to 223 officers and men due to the increased automation of formerly semi-automatic systems.

It had been found in practice that the use of compensated fuel tanks was unsuccessful because of the growth of marine organisms at the oil-water interface which reduced the amount of usable fuel. The consequence was that the intended range of the class as built was rarely achieved. However, as a result of the reduction in topweight of the Batch 2 Exocet conversion over the original design, plus the addition of permanent ballast, it was possible to revert to the separate ballast tank system. The result was a marked increase in the amount of fuel that could be consumed and a commensurate increase in the endurance achieved.

At the end of her refit the standard displacement of CLEOPATRA had increased to 2,700 tons, with a full load displacement of 3,200 tons. Her mean keel draught had increased to 14.75 ft, with navigational draught to propeller tips now nearly 19 ft. Top speed was reduced by about two knots. CLEOPATRA recommissioned on Trafalgar Day 1975 and on completion of trials became leader of the Fourth Frigate Squadron but at this time did not operate a Lynx helicopter. PHOEBE started her refit in August 1974 at Devonport and was due to complete at the end of October 1976. However, the ship did not recommission until 29th April 1977, as half leader of the First Frigate Squadron, and one of her first tasks was to participate in the Royal Silver Jubilee Fleet Review in June that year. In February 1978 PHOEBE became the first frigate to embark a Lynx helicopter, over two years after CLEOPATRA has recommissioned.

It is interesting to note that at more or less the same time that ARIADNE was being built by Yarrow as the last British Leander, the same company were building two "Broad-beam" Leanders for the Chilean Navy. However, the builders had modified the design to incorporate several features that were to appear in the Batch 2 conversion. Thus, while retaining the twin 4.5 in guns, the Chilean ships were also fitted with Exocet missiles, the four single launchers, angled outwards and forwards, being mounted right aft on the quarterdeck. In addition the extended shelter deck seen in CLEOPATRA and PHOEBE was fitted with US Mk 32 ASW torpedo tubes while the Mk 10 mortar was omitted. ALMIRANTE LYNCH and ALMIRANTE CONDELL, the Chilean ships, were completed in October 1974, before CLEOPATRA completed her refit. Thus, the Chilean ships had the same anti-ship missile armament as the Royal Navy's Batch 2 conversions but managed to retain a medium calibre gun system. This raised the question as to why the British Leanders did not adopt the Chilean arrangement and so retain a far greater general purpose capability. No satisfactory answer has ever been forthcoming, but one can imagine the Royal Navy arguing that as their ships were going to operate in task groups they could afford to be more specialised, whereas the Chilean ships were intended for more independent action and so needed to have greater versatility. The fact that the Royal Navy had been premature in discarding the medium calibre gun was suggested once again when the Dutch Leanders were refitted, from 1976 onwards, with the modern 76 mm DP gun, as well as the American Harpoon missile system.

SIRIUS was the third ship to be refitted, again at Devonport, starting in March 1975. Compared with the previous conversions, she was to be slightly modified, having the STWS-1 tubes mounted on the main deck, instead of the shelter deck. In order to accommodate the system on the main deck, part of the outer wall of the hangar was removed to allow the mount to train, as well as permitting reloading from the magazine below the hangar. SIRIUS recommissioned in February 1978.

The refit of MINERVA at Chatham was a very protracted affair, the ship paying off in October 1975 following a deployment to the West Indies. Whilst on post refit trials in October 1978, MINERVA suffered a boiler room explosion at Portsmouth and had to be towed back to Chatham for repairs. At long last, she eventually recommissioned in March 1979. Then, in December of that year, while alongside at Devonport, a dockyard crane fell on the ship during a storm, destroying the starboard Seacat launcher, damaging the hangar and several communications aerials.

ARGONAUT, which started her refit in February 1976 and recommissioned in March 1980, was the only Leander actively involved in the Falklands campaign. The ship suffered badly in Argentine air attacks on 21st May 1982 in San Carlos Water. In an early attack, her Type 965 radar antenna and some upper deck fittings were damaged. Then, at 1730, she was attacked by six A-4 Skyhawks and was hit by two bombs which did not explode. One entered ARGONAUT just above the waterline at the bulkhead between the boiler and engine rooms, smashing steam pipes in both compartments, as well as disabling the main air-blower and causing one of the two boilers to rupture. The second bomb hit ARGONAUT five feet below the waterline, passed through two fuel tanks into the magazine, holing the starboard side as well. The magazine was devastated as two Seacat missiles and other ammunition exploded, blowing off the door in the upper deck, lifting the deck of the forward seamen's mess by four feet, destroying cable runs and causing fires. Two seamen who were putting a missile into the hoist were killed instantly, but the magazine fire and explosions were extinguished by an inrush of diesel oil from the ruptured fuel tanks. ARGONAUT lost steering and was heading at full speed for Fanning Head until the Forecastle Officer let go an anchor which brought the ship to a stop just before she hit the headland. The frigate spent another eight days in "Bomb Alley" before limping back for repairs to Devonport, where she arrived having taken a month to make the 8,000 mile journey. While in dockyard hands, ARGONAUT underwent an extensive refit which included the installation of the towed array sonar system.

MINERVA was due to undergo a similar towed array refit and had actually paid off on 2nd April 1982 but, because of the invasion of the Falklands, the ship was kept running despite continued bouts of engine trouble. She had been scheduled to return to Falklands waters with the destroyer LIVERPOOL in late 1982, but because of persistent machinery troubles was sent to Portland, where on 22nd November, she rammed YARMOUTH's stern, while berthing. Eventually, in August 1983, MINERVA started a seven and a half month, £8,000,000 refit at Devonport, during which 16 additions were made, including 100 tons of extra ballast, new communications equipment and two additional 20 mm guns, one right aft and one on the former port boat deck. The refit was completed on 2nd March 1984 and MINERVA recommissioned fourteen days later. Subsequently, she was deployed to the South Atlantic on Falkland Islands patrol from October 1984 until March 1985.

HMS MINERVA leaving Portsmouth 16/5/85. Note single GAM B01 20 mm gun on the starboard side of the quarterdeck
L. van Ginderen

HMS DANAE at Portsmouth 18/5/83   L. van Ginderen

DANAE started her Exocet conversion in August 1977 at Devonport and completed in April 1981. She had been built with a VDS well but the sonar was never fitted and the well had been plated-in during an earlier refit. Less than three years after completing her conversion, DANAE underwent a six week assisted maintenance period at Devonport to improve her close-range armament as a result of lessons learned during the battle for the Falklands. Thus, she emerged from Devonport on 19th July 1983 with a twin 20 mm Oerlikon mount fitted aft, two B-MARC GAM BO1 20 mm guns in lieu of the ship's boats, and mountings for up to eight general purpose machine guns. In addition a new RAS winch was fitted close to the funnel and two sextuple launchers for American Super Rapid Blooming Off-Board Chaff (SRBOC) were fitted on the superstructure. Since the Falklands war, DANAE has made several deployments to the South Atlantic for patrol duties, the most recent being a five month trip completed in September 1988.

However, this last tour must have had a deleterious effect on the old ship's hull because, when she entered Devonport in early 1989 for a twelve week docking period, a survey highlighted the need for major repairs to the engine and boiler rooms. The survey also revealed that the steel hull plating of the engine room compartment did not meet the minimum thickness requirements, while in the boiler room many frames and longitudinals had to be replaced in addition to the shell plating. Accordingly, DANAE's docking period increased to fifteen months and the frigate started post refit trials in the Spring of 1990 before starting a five month deployment to the South Atlantic.

The last ship of this group to refit was PENELOPE, which had spent the previous ten years as a trials ship until she paid off in December 1977. All Seawolf trials equipment was removed and the VDS well plated over. Refitted with Exocet, STWS-1, three Seacat launchers and an enlarged hangar and flight deck for a Lynx helicopter, PENELOPE recommissioned on 22nd January 1982 after four years in dockyard hands. Within a few months she had been equipped with additional close-in weapons. Thus, at the stern and just to port of the centreline, a B-MARC GAM BO1 20 mm AA gun was mounted on a small bandstand, while additional chaff launchers were fitted on the quarterdeck. These SRBOC projectors were angled forward and slightly to starboard and intended to fire either chaff or Infra-red cartridges to decoy incoming anti-ship missiles, more rapidly than Corvus. The SRBOC system was developed by the American Hycor Corporation as an instant-reaction decoy system. Canisters of chaff are fired from the six-barrelled launcher to provide a protective envelope of chaff sufficient to act as a decoy and so seduce the missile away from its real target.

HMS PENELOPE in Falklands waters in 3/84   L. van Ginderen

PENELOPE was particularly busily employed in the South Atlantic during 1983-4, making three complete deployments in just eighteen months. In July 1985 the ship started yet another tour of service in the rough waters of the South Atlantic.

On return to Devonport PENELOPE was taken in hand in January 1986 for an intended thirty four-week refit which included the re-tubing of her boilers and radar updating. Following post refit trials the ship was rededicated on 25th September before visiting Portland to work-up. Yet another trip to the South Atlantic followed in May 1987 when PENELOPE replaced MINERVA at the end of the latter's five months on patrol off the Falklands.

Apart from deploying south the three Exocet Leanders have been extensively involved in British and NATO exercises, MINERVA for example taking part in Exercise Purple Warrior, a large amphibious exercise off the west coast of Scotland in November 1987. PENELOPE participated in Teamwork '88 in the North Sea and Atlantic during October 1988, unfortunately colliding with the Canadian supply ship PRESERVER during RAS operations on 11th September 1988 and coming off second best. At first the damage was reported as light but later reports suggested that underwater damage had been extensive, two repair teams of twenty divers from USS PUGET SOUND undertaking twenty three hours of underwater welding to keep the ship afloat. PENELOPE subsequently returned to the UK on 28th October for docking and repairs at Thew Engineering, Southampton. It was expected that this would take between eight to ten weeks but the ship was still in dock during February 1989. PENELOPE worked-up at Portland from mid-April 1989 to the end of May before commencing her sixth deployment to the South Atlantic since 1982. In December 1989, while still in Falklands waters, PENELOPE, then the oldest major unit in the Fleet, clocked-up 500,000 miles since commissioning. The ship returned to the UK in April 1990 after visits to South America, the Caribbean and Florida. In August of that year it was announced that PENELOPE and MINERVA would pay off in the following December.

In 1980 CLEOPATRA was selected to receive the Type 2031 towed array sonar system (TASS). Subsequently, in December of the following year the frigate paid off to be fitted with this sonar system which, because of its weight and volume, was to cause significant changes to her superstructure and weapons fit, as detailed in the following chapter.

HMS PENELOPE, at Antwerp 8/88. She was the last of seven ships to receive the Batch 2 conversion
L. van Ginderen

CHAPTER FIVE

BATCH 2A TOWED ARRAY SONAR CONVERSIONS

In June 1982 the Batch 2 frigate PHOEBE, which had completed an Exocet conversion in April 1977, emerged from a refit at Chatham showing a number of significant changes. The forward Seacat launcher and its associated director on the bridge, the Type 965 radar on the mainmast, and the seaboats and their davits had been removed. The superstructure had been reduced by cutting away a considerable amount of the structure abaft the foremast and forward of the funnel and, in addition, the mainmast had been replaced by a new, smaller mast with Type 1010 IFF at its head. The single 40 mm guns abreast the bridge had been replaced by two 20 mm Oerlikon guns and two additional 20 mm guns, one above the bridge and one aft, were installed while, the two sets of triple STWS-1 ASW torpedo tubes had been resited on either side of the hangar on the upper deck and the Exocet launchers were lowered. All in all a determined attempt seemed to have been made to reduce topweight so as to allow the installation of a towed array sonar.

HMS PHOEBE leaving Portsmouth 2/7/82 on completion of work to prepare her to receive Type 2031 "I" TASS
L. van Ginderen

Towed array is a passive sonar system on cables towed 1,000 yards astern of a surface ship or submarine and is capable of detecting submarines at ranges in excess of 100 miles. By towing the sonar far astern, the noise generated by the towing ship causes far less interference, thereby allowing submarines to be detected at very much greater distance than hitherto considered normal. An experimental towed array sonar was tested aboard the Rothesay class frigate LOWESTOFT for several years and, upon successful conclusion of these trials, the decision was made to install the system in operational frigates. The ships selected were the Batch 2 Exocet-armed Leanders PHOEBE, CLEOPATRA, SIRIUS and ARGONAUT. However, when PHOEBE left Chatham in June 1982, the towed array sonar, now designated Type 2031, had not been installed, although the ship had been prepared to receive the equipment.

The Type 2031 'I' TASS (Towed Array Sonar System) installed in these four ships was built to a specification put forward by the Admiralty Research Establishment at Portland. The project was led by Marconi Avionics, who supplied the signal processing electronics, while Ameeco Hydrospace (now part of Plessey) provided the towed, liquid-filled instrumentation module and its highly sensitive hydrophones, with Clarke Chapman supplying the winch gear. It is understood that only five Type 2031'I' sets were built and the system has been developed into the Type 2031'Z' which is being fitted to Type 22 and 23 frigates. Although few particulars have been released about the Type 2031 TASS quite a lot has been made available about its export version, the Plessey COMTASS passive search sonar. COMTASS, which has an overall capability reported to be slightly less than that of the Type 2031, is normally operated at 5-15 knots but can be towed at 30 knots. The towed array consists of a 1,000-1,500 m length of 25 mm diameter towing wire connected to a 63 mm diameter array which is 82 m in length. The array is made up of a gel-filled vibration isolation module (VIM), a liquid-filled instrumentation module with the acoustic aperture array module of highly sensitive hydrophones providing a beam pattern for detecting low-frequency noises, and a second VIM. The array is deployed and recovered by winch gear on the stern and the data processed by a range of micro-processors in the sonar operations room. During ASW operations, TASS is used in conjunction with shipborne active/passive sonars augmented by sonar inputs, via data links from other ships and aircraft in the area.

HMS CLEOPATRA entering Portsmouth 10/8/84. She was the first Leander to be fitted with TASS
L. van Ginderen

On 10th June 1983, CLEOPATRA, the first Batch 2 Leander to be equipped with towed array sonar, emerged from Devonport Dockyard with modifications identical to those effected in PHOEBE, plus the addition of a massive cable reel and other associated gear at the after end of the flightdeck. This conversion involved the addition of 70 tons of extra topweight which goes a long way to explaining the obvious attempt to save topweight elsewhere when installing this system. Unfortunately, the original Royal Navy design for the installation of the Type 2031'I' sonar system involved considerable cramping of the flight deck and also filled the helicopter hangar with the electronics display room necessary for operation of this sonar. Consequently, CLEOPATRA lost her ability to operate a Lynx helicopter.

HMS CLEOPATRA leaving Portsmouth 8/12/86. Note winch and cable drum on the centreline of the quarterdeck
L. van Ginderen

This was clearly an unacceptable state of affairs and subsequent installations were modified to permit the operation of a helicopter. This alteration involved the construction of a sponson on the corner of the starboard quarter so as to move the cable reel and associated gear away from the operating area of the flight deck. The towed array sonar display room was relocated in the former forward Seacat magazine, which had been left empty after the launcher had been removed. SIRIUS, which completed her TASS refit in Devonport in September 1983, was the first of the class to be given this modified installation. She was followed into service by ARGONAUT which completed a sixty eight week period of refit and repair at Devonport on 18th November 1983. ARGONAUT had been badly damaged by the Argentine Air Force in May 1982 and while she was under repair the opportunity

HMS PHOEBE 9/7/88. The winch and cable drum for Type 2031 'I' TASS can be seen on a platform built out from the starboard quarter
L. van Ginderen

HMS SIRIUS, arriving at Antwerp, 27/6/87, equipped with TASS   L. van Ginderen

was taken to install the towed array sonar gear, originally intended for MINERVA, on a sponson on the starboard quarter. Finally, on 20th January 1984, PHOEBE completed a four month refit at Devonport during which the Type 2031'I' sonar was installed finally and the single 20 mm guns on the bridge and quarterdeck removed.

The 1981 Defence White Paper stated that major refit costs had increased to such an extent that those ships at that time unmodernised would pay off early and not be improved to the standard of their modernised sisters. However, it was found possible to make some major modifications during short docking periods and this may account for the partial reconstruction of PHOEBE in 1982, which was completed during a docking period in 1983. It may well be that the Royal Navy had found a way of carrying out quite substantial updating of its ships during routine docking periods and so circumvented one of the more ill-thought-out parts of the unfortunate 1981 Defence Review. An alternative, if cynical, explanation was that dockyards under threat of closure had found a way of working more efficiently.

Since being fitted with Type 2031'I' these four ships have been engaged in trials and exercises which have shown conclusively the value of the TASS. As a result, the far superior Type 2031'Z' system has been fitted in Batch 2 and 3 Type 22 frigates and is likely to be installed in the earlier Batch 1 Type 22 vessels, as well as providing a major detection system in the Type 23 Duke class frigates now under construction. However, prolonged use of such sonars aboard Leander class frigates resulted in metal fatigue problems and consequently, in 1989 it was announced that the appropriate part of the Type 23 hull will be strengthened to prevent similar problems occurring in these new ships.

HMS ARGONAUT leaving Portsmouth 17/12/84   L. van Ginderen

In summer 1984 ARGONAUT suffered a furnace explosion necessitating a major rebuilding of her starboard boiler at Devonport. These costly repairs were completed in October of that year. A few months later in spring of 1985 her sister SIRIUS entered Devonport for a routine refit, while PHOEBE was in dockyard hands in late 1986 to early 1987. Surprisingly, when CLEOPATRA underwent a major refit in Devonport Dockyard from May 1987 to December 1989, she was not brought up to the same standard as SIRIUS and therefore remains unable to operate a helicopter.

SIRIUS made an unscheduled entry into Devonport's floating dock on 7th March 1988 for extensive and unprogrammed structural repairs before joining the "Outback '88" deployment to Australasia and the Far East.

HMS SIRIUS 20/10/88   M. R. Dippy

On 28th March ARGONAUT followed CLEOPATRA into dockyard hands for an extensive refit which was completed on 15th October 1989. Finally, SIRIUS started a forty week long refit at Devonport on 23rd October Clearly, these Leander class frigates, which have seen twenty years hard service are beginning to show their age and will require expensive regular maintenance to keep then operational in the early 1990s. PHOEBE, however, was an early victim of the "peace dividend" brought about by the ending of the "Cold War" during 1990. Thus, in August it was anounced that PHOEBE would be paid off for disposal in December 1990 as part of the Defence cuts resulting from the rapprochement between East and West.

CHAPTER SIX

BATCH 3 SEAWOLF CONVERSIONS

The last ten Leanders built for the Royal Navy were given an additional 2 feet more beam so as to increase their stability and endurance as well as allowing greater scope for future modernisations. As built, the ten ships of the Andromeda class were identical to the original design in terms of equipment, but because of their greater internal volume they could be reconstructed to carry the new Type 2016 passive search sonar and the Seawolf point defence missile system both of which were due to enter in 1979 aboard the frigate BROADSWORD, the first of the Type 22 frigates. The Type 22s were designed to operate in the Greenland-Iceland-United Kingdom (GIUK) Gap using the Type 2016 sonar to detect Soviet submarines at very long range and then, using one of their two Lynx helicopters, destroy the submarine(s) with homing torpedoes. As they would be operating far from the support of shore-based aircraft and were most likely to be attacked by submarine or aircraft launched missiles, the Type 22s were equipped with two sextuple launchers for the Seawolf anti-aircraft/anti-missile missile system. Finally, to counter any attack by missile-armed surface ships, the Type 22s were equipped with four Exocet sea-skimming surface-to-surface missiles.

Accordingly, the objectives of the modernisation effected in the Batch 3 Leanders were to redress the fighting power lost since their completion, to replace systems which had become obsolete or proven unsatisfactory and to reduce overall running cost by standardisation on equipment being fitted in the Type 22 frigates then under construction. However, as the Type 22 design was for a ship some 1,300 tons heavier, 58 feet longer and with 5.5 feet more beam than a Batch 3 Leander, it was clear that not all the features of the new design could be built into ANDROMEDA and her sisters. Fortunately, because of that extra two feet of beam and by means of an almost complete internal rearrangement, which included resiting of some of the bulkheads together with the elimination of every possible amount of unnecessary topweight, it proved possible to incorporate the key features of the Type 22 design into the smaller Leander hull.

The reconstruction involved the removal of all obsolescent equipment including the twin 4.5 in gun turret, the Seacat missile system, the Mk 10 AS mortar together with all radars and fire control systems. The mortar and VDS wells were plated-in to provide more accommodation and an increased flight deck area. The quarterdeck was cut away at the corner of the port quarter to provide space at main deck level for the small hydraulic crane, originally sited at the after end of the flight deck, used for streaming towed decoys and other equipment. The superstructure was reduced markedly by removal of deckhouses around the base of the mainmast, elimination of several platforms from the hangar sides and by cutting away a considerable amount of superstructure between the funnel and the after end of the base of the foremast. Finally, even the domed top to the funnel was removed in an effort to save unnecessary topweight although the hangar had to be enlarged to take a Lynx helicopter.

HMS JUPITER, 19/7/86. The funnel, after superstructure and mainmast have been cut down to reduce topweight
L. van Ginderen

As reconstructed, the Batch 3 Leanders are equipped with a hand-loaded sextuple Seawolf launcher mounted on a platform at the break of the forecastle and controlled by a Type 910 tracking radar with its associated command aerials mounted above and abaft the bridge. The foremast was enlarged to provide a broad platform at its head for the new ESM equipment and a fully stabilised Type 967/968 radar aerial which provides warning/target indicating data for the Seawolf GWS 25 missile system. Type 967 is a L-band pulse-doppler air search radar and is mounted in a back-to-back arrangement with a Type 968 S-band radar which provides low level air warning cover and surface search. The combined Type 967/968 aerial rotates at 30 rpm and because it is fully stabilised remains parallel to the horizon when the ship heels over. A Type 1006 navigation radar is mounted on a lightweight platform projecting forward and to port halfway up the foremast while the usual array of communications gear is fitted at the crosstrees at the head of the foremast.

HMS HERMIONE at Portsmouth 13/1/86. The back-to-back Type 967/968 radar aerial can be seen at the head of the foremast   *L. van Ginderen*

Four containers for Exocet surface-to-surface missiles are mounted directly on the forecastle deck in pairs angled out to port and starboard. This arrangement is far superior to that of the Batch 2 conversions where the height of the platform on which the Exocet canisters were mounted was such that it restricted severely the field of fire of the forward Seacat launcher. By comparison, the Seawolf launcher in the Batch 3 conversion has a far greater field of fire as it is only "wooded" by the superstructure and masts abaft the Exocet launchers. The armament is completed by two single 20 mm guns in a new position abreast the foremast and one deck lower than before, at boat deck level, and two sets of Plessey STWS-1 triple torpedo tubes mounted at upper deck level abaft the hangar. The pair of BBC/Corvus chaff launchers were retained in their position at boat deck level, abreast a new, much taller and thinner mainmast which supports a variety of electronic warfare equipment and is surmounted by a pole aerial. New lightweight platforms for SCOT radomes were installed abreast the mainmast, just forward of the chaff launchers. Another obvious change effected during this radical conversion was the removal of all ships' boats and davits, and their replacement by a pair of Avon Seariders with a new lightweight handling crane.

Amongst the most significant changes effected during this reconstruction was the installation of the new Type 2016 long range sonar in lieu of the earlier types fitted during construction, although a Type 162M bottom search was retained. These sonar sets together with the various radars plus ESM, ECM and ECCM equipment provide information for an automatic data handling system (ADAWS) which then evaluates the threat and automatically initiates an appropriate response. It is understood that this version of ADAWS is made up of five Ferranti 1600 series computers and because of the increased electrical demand the generating capacity has been increased to 2,500 kW. In addition, JUPITER and SCYLLA were fitted with Rationalised Internal Communications Equipment (RICE). The use of automation wherever possible has reduced the ships' complement from 260 to just 214, a significant saving in terms of operating costs and future demographic trends which suggest that there will be fewer men of age available for military service from the late 1980s — early 1990s.

The first Batch 3 Leander to undergo this modernisation was ANDROMEDA which paid off at Devonport in January 1978 and emerged three years later with an ASW and self-defence capability matched only by that of

HMS ANDROMEDA, 15/7/89. Note single 20 mm GAM B01 gun on a bandstand on the starboard side of the quarterdeck and the port side cut out in the hull for the towed decoy derrick
L. van Ginderen

HMS CHARYBDIS entering Portsmouth 19/7/87. The sextuple Seawolf launcher can be seen forward of the Exocet launchers, ahead of the bridge   *L. van Ginderen*

the brand new Type 22 frigates. Subsequently, CHARYBDIS completed her refit at Devonport in July 1982 and was followed by HERMIONE which left Chatham on 21st June 1983. However, the latter's refit had not been "finished off" as she spent the period from mid July 1983 to January 1984 at Devonport. JUPITER returned to service on completion of her conversion at Devonport on 15th October 1983, leaving only SCYLLA under conversion. Work on SCYLLA was delayed by the Falklands war of April to June 1982 which diverted men fitting out SCYLLA to more urgent tasks such as fitting out requisitioned merchant ships and repairing battle damage. Furthermore, it is understood that SCYLLA was cannibalised to some extent to provide urgently needed spares and these two factors combined to delay the completion of her modernisation until 7th December 1984.

Since the completion of this modernisation at least one of the class has been fitted with additional guns and chaff launchers for service in the South Atlantic. Thus, CHARYBDIS was observed in July 1983 fitted with three additional 20 mm B-MARC GAM BO1 guns, one on the quarterdeck and two abreast the mainmast, plus two six-barrelled SRBOC chaff launchers just forward of the new 20 mm guns.

Originally it had been intended that all ten Andromeda class ships would receive this elaborate and expensive reconstruction at an estimated cost of £70,000,000 per ship. However, the Defence Review of June 1981 caused the abandonment of further Leander modernisations and as a result, BACCHANTE was sold to New Zealand while the long term future of APOLLO, ACHILLES, ARIADNE and DIOMEDE remained unclear. Subsequently, all four ships were refitted for further service, but no attempt was made to update them and consequently, after 1986, were employed mainly as training ships.

Since completing her reconstruction in 1981, ANDROMEDA has been employed extensively on duties in the Persian Gulf and the South Atlantic. Thus, she joined the Task Force off the Falkland Islands in early June 1982 before the end of the war, making another deployment to the area in the spring-summer of 1984. ANDROMEDA was on patrol in the Persian Gulf/Indian Ocean in late 1983 joining the "Orient Express" deployment involving INVINCIBLE, ACHILLES and AURORA, as well as undertaking a second tour of duty in the area during March-June 1985. On 2nd September 1985, ANDROMEDA paid off at Devonport at the start of her first refit since recommissioning in 1981, having steamed 205,000 miles and spent six hundred and fifty five days at sea.

Of her sisters, CHARYBDIS made one deployment to the South Atlantic in early 1983 and two to the Gulf area, one in spring 1984 and the second in the summer of 1985 while HERMIONE served in the Persian Gulf

HMS HERMIONE at Portsmouth 23/2/85. Note the temporary deckhouse, with two aerials on the roof, fitted on the flight deck during trials of the Guardian ESM system   *L. van Ginderen*

in late 1984. In the summer of 1985 HERMIONE, by then fitted with a 20 mm GAM B01 gun on her starboard quarter right aft, carried out trials with the Thorn-EMI "Guardian" EW system. During these trials the frigate was fitted with a temporary deckhouse, with two "Guardian" antennae on its roof, on the flight deck immediately aft of the hangar. In December 1983, JUPITER took part in trials in Cardigan Bay aimed at demonstrating the effectiveness of Seawolf against incoming anti-ship missiles. This involved JUPITER firing an Exocet over the horizon with the Type 22 frigate BRILLIANT positioned downrange offset from the line of fire — an exacting target for the Seawolf system. BRILLIANT acquired the Exocet under normal conditions and then fired one Seawolf missile in the preferred television mode for low level operations, successfully destroying the Exocet at long range. Finally, SCYLLA, which had only recommissioned in December 1984, spent the first eight months of 1985 involved in trials, training and work-up prior to becoming a fully operational ship. She then took part in exercise "Autumn Train" in October of that year.

HMS SCYLLA, 3/5/85. She was the last of the Seawolf conversions   *L. van Ginderen*

In May 1987, SCYLLA took part in trials with the first production Ferranti Type 2050 sonar set. The only visible feature was the installation of a large grey/green "box" added immediately aft of the Seawolf launcher which prevented the latter from being either loaded or trained. The "box", which would also have prevented firing of the Exocet missiles, had been removed by mid June of that year.

All five ships were used regularly as members of squadrons sent to the Persian Gulf area for Armilla patrol duty. This involved accompanying British-registered ships through the Straits of Hormuz during the period from November 1986 to August 1988 at the height of the war between Iran and Iraq and exposed the frigates to the risk of attack from a variety of weapons. In an attempt to reduce their vulnerability to attack by anti-ship missiles, all five frigates were fitted with radar suppression material. This consisted of strips of grey-painted vinyl sheeting stuck to all upperwork surfaces, including ready-to-use lockers and Exocet missiles launchers, to reduce their radar signatures. In addition, the ships were fitted with two 18-barrelled Wallop Barricade chaff launchers on the after corners of the hangar roof which seems to be the largest "radar target" on a Leander. The Wallop Barricade chaff launchers were used to supplement the decoy capability of the Corvus and SRBOC chaff launchers already installed.

As reconstructed, these five Batch 3 Leanders have most of the capabilities of the much larger Type 22 frigates of the Broadsword class and consequently they represent a valuable and potent addition to the Royal Navy's escort forces. In view of their relative youth, when compared to the rest of the class, and the completeness of their reconstruction, it is to be hoped that they will be capable or remaining in service until the mid 1990s — they will be needed until then at least because of the very slow rate of ordering of Type 23 Duke class frigates.

CHAPTER SEVEN

OVERSEAS LEANDERS

The Type 12s proved to be popular and successful ships with the Royal Navy and not unnaturally attracted overseas customers including the navies of Australia, India, New Zealand and South Africa. Consequently, when the ships of the modified Type 12 or Leander class started to join the Royal Navy in the early 1960s, they were also the subject of overseas interest. For example, there are reports that Argentina and Spain were interested in acquiring Leander class frigates although, in the event, no orders were forthcoming. The Spanish had been negotiating for the construction of four Leander class ships but, unfavourable British Parliamentary and Press comment about the proposed deal and the Spanish regime led to the collapse of discussions following the election of a Labour Government in 1964. Despite these setbacks, Leander class frigates entered service with the navies of Australia, Chile, Holland, India and New Zealand.

AUSTRALIAN RIVER CLASS FRIGATES: SWAN AND TORRENS

In August 1950 the Royal Australian Navy announced plans for six Type 12 ASW frigates but subsequently, the number of ships was cut to four. Some years later, the fifth and sixth ships were reordered to a modified design which incorporated many of the features of the Royal Navy's Leander class frigates.

HMAS PARRAMATTA, 27/6/86, an Australian Type 12 frigate   M. R. Dippy

The Australian Type 12 River class frigates form three distinct groups with YARRA and PARRAMATTA, which were completed in July 1961, being basically the Royal Navy's Rothesay class with the addition of a Dutch LW02 radar on the tubular foremast. By comparison, the second group — STUART and DERWENT — which were completed in 1963-64, combine the Rothesay hull and some Leander modifications with, for example, the main deck extended right aft to the quarterdeck on the port side only with an Ikara AS missile launcher on the starboard side. The third group, SWAN and TORRENS, were laid down on 18th August 1965 to an extensively updated Rothesay design modified to Leander standards but without the helicopter facilities.

Thus, in this pair of ships the forecastle was extended right aft and a new continuous superstructure with a streamlined funnel replaced the former light superstructure found in the Rothesay class. The armament consisted of an Australian-built twin Mk 6 4.5 in gun turret forward of the bridge, a quadruple Seacat surface-to-air missile launcher on the aft superstructure and a Mk 10 AS mortar in a well on the quarterdeck. The armament was completed by a single launcher for Ikara AS missiles mounted in an aperture on the starboard quarter. The electronic suite consisted of a LW02 air surveillance radar on a stump mainmast, a Type 8GR-301 surface/navigation aerial on the foremast, a M22 FCS at the head of that mast plus a M44 Seacat fire control radar mounted on the centreline aft of the mainmast with an Ikara FCS mounted on the bridge. The sonar outfit consisted of Types 162, 170 and 177M.

HMAS SWAN, 5/5/87   *M. R. Dippy*

SWAN was launched by Williamstown Naval Dockyard, Melbourne, on 16th December 1967 while her sister TORRENS was launched on 28th September, 1968 from Cockatoo Dock and Engineering Ltd., Sydney. Both ships commissioned within a year of each other, SWAN entering service on 20th January 1970 and her sister on 19th January 1971. As completed they displaced 2,100/2,700 tons and measured 370 feet overall with a beam of 41 feet and a draught, at the propellers, of 17.25 feet. Like their earlier sisters they are propelled by two double reduction geared steam turbines which develop 30,000 shp and drive two shafts to give a top speed of 30 knots with a range of 3,400 miles at 12 knots. Their complement consists of 13 officers and 234 men.

All in all, SWAN and TORRENS are very similar in appearance and performance to their British Leander class cousins despite the absence of an ASW helicopter such as the Wasp. However, this defect is offset to some degree by the provision of an Ikara long range AS missile system which, although of shorter range than the helicopter, is capable of use in all weathers. In the early 1980s SWAN and TORRENS were scheduled to have been given 18 month refits to enhance their ASW capability by installation of the Australian-developed Mulloka sonar, two sets of triple ASW torpedo tubes and improved electronics.

The hull, machinery and superstructure were to be thoroughly overhauled and the Mk 10 AS mortar removed. Unfortunately, lack of development of the Ikara system and the limitations of the ageing Seacat system meant that the value of this modernisation would be limited. In the event, because of the need to maintain the operational strength of the fleet, these refits were delayed to such an extent that SWAN was scheduled to enter dockyard hands in December 1983 and TORRENS in September 1984. Subsequently, a change of Government leading to a re-evaluation of naval priorities including the value of the proposed modernisations of SWAN and TORRENS led to the cancellation of these refits in late 1983. In effect, both ships were considered to be not worth reconstructing and SWAN started a year-long refit in December 1983. While in dockyard hands she was equipped with the Mulloka sonar, Mk 32 torpedo tubes and a torpedo decoy system. TORRENS underwent a similar refit during the period from September 1984 to September 1985.

HMAS TORRENS 9/11/88. Note hull cut-away in starboard quarter to house the Ikara AS missile launcher
      M. R. Dippy

CHILEAN LEANDERS: ALMIRANTE CONDELL AND ALMIRANTE LYNCH

In 1970 the Chilean Navy decided to purchase two Leander class frigates from Yarrow & Co Ltd., Scotstoun as part of their modernisation programme. Both ships were to be of the "Broad-beam" Andromeda class with a displacement of 2,500/2,962 tons, measuring 372 feet overall with a beam of 43 feet. As with their British counterparts, propulsion is by means of two sets of geared turbines which develop 30,000 shp to give a top speed of 29 knots and a range of 4,500 miles at 12 knots. ALMIRANTE CONDELL, which originally was to have been named ALMIRANTE LATORRE until this name was allocated to the then-newly purchased former Swedish cruiser GOTA LEJON, was laid down at Scotstoun on 5th June 1971, launched on 12th June 1972 and commissioned on 21st December 1973. Her sister ALMIRANTE LYNCH was laid down on 6 December 1971, launched exactly twelve months later and commissioned on 25th May 1974. Contrary to popular belief, neither ship was delayed by either Governmental or labour interference, which was threatened following the overthrow of the Allende regime. Indeed, ALMIRANTE LYNCH was delivered several days ahead of schedule in order to prevent such interference, which did not materialise. Subsequently, ALMIRANTE CONDELL and ALMIRANTE LYNCH worked-up in British waters in 1974 but both frigates were in Chilean waters by February 1975.

The Chilean Leander ALMIRANTE CONDELL. She is equipped with Type 992Q target indicating radar at the foremast head, Mk 32 triple torpedo tubes to port and starboard abreast the hangar and four Exocet missile container/launchers on the quarterdeck   *Chilean Navy*

Their armament consisted of the usual twin 4.5 in DP gun turret forward of the superstructure with its MRS3 FCS on the bridge plus a quadruple Seacat launcher to port, on top of the hangar with its controlling GWS22 director to starboard. As with the Andromeda class, two single 20 mm guns were fitted to port and starboard abreast the foremast but neither VDS nor Mk 10 mortar were installed. Instead, four Exocet surface-to-surface missiles were placed on the quarterdeck, as opposed to the British Batch 2 conversion in which the missiles replaced the 4.5 in gun turret thereby reducing versatility. The armament of the Chilean Leanders was completed by two sets of triple Mk 32 ASW torpedo tubes, mounted on a shelter deck abreast the hangar, and a light helicopter. A Type 965, with AKE-1 array, long range early warning radar aerial is mounted on top of the mainmast with a Type 1006 navigation radar mounted on a platform halfway up the foremast. Compared to the British Leanders, ALMIRANTE CONDELL and ALMIRANTE LYNCH have markedly taller foremasts with a Type 992Q surface search/target indicating radar aerial at the masthead — the only Leanders to be thus equipped. The sonar suite consisted of a Type 162 side-looking classification set, a Type 170 hull-mounted 'searchlight' with a range of 2,500 m and a Plessey PMS-32 medium range (6,000 m) search sonar.

These ships, which have a complement of 263 officers and men, have proven very successful in service. Despite the great distance between the United Kingdom and Chile, and no doubt because of the excellent technical skills in the Chilean Naval Dockyards (Astilleros y Maestranza de la Armade — ASMAR), it has been possible to maintain these frigates in excellent condition. It is understood that one of the pair has been given a long refit/modernisation at Talcahuano but no details of any changes are available at the time of writing. There were rumours in 1983/84 that the Chilean Navy would like to purchase two ex-Royal Navy Leanders to create a four ship Leander division. Subsequently, there were reports that the Chileans had hopes of acquiring LEANDER when she paid off but the deal fell through because the Australians would only sell the Ikara missile system to Commonwealth nations.

DUTCH LEANDERS: VAN SPEIJK CLASS

Early in the 1960s the Dutch Navy was allocated funds to build six frigates to replace six ex-USN destroyer escorts of Second World War vintage. Because the Dutch Government insisted that the money was spent immediately, the Navy had to adopt an existing design and the British Leander design was judged to be suitable and was available for purchase. However, the design had to be re-worked into metric measurements for construction and modified to accommodate Dutch equipment wherever possible. Despite these problems, the first four ships of the Van Speijk class were ordered in 1964 and all six frigates were operational by mid 1968.

HNLMS EVERTSEN, 12/77, a typical Dutch Leander as completed   Mike Lennon

As completed, ships of the Van Speijk class displaced 2,200/2,835 tons, measured 113.6 × 12.5 × 4.6 m and were powered by geared steam turbines giving a top speed of 28 knots and a range of 4,500 miles at 12 knots. In essence they were typical Leander class frigates with the same hull form, superstructure and weapons systems as their British counterparts. Thus, the Van Speijks were armed with a twin Mk 6 4.5 in gun turret forward of the bridge, two launchers for Seacat surface-to-air missiles on the hangar roof, a Mk 10 mortar in a well aft of the flight deck and operated a Wasp helicopter. However, their electronic equipment, such as radar, sonar and fire control systems were largely Dutch and this difference in antenna shape provided an obvious recognition feature. The Van Speijk class frigates, like their British half-sisters proved to be reliable and effective ships in service but their design pre-dated the introduction of anti-ship missiles and consequently they became increasingly underarmed in comparison with more modern designs.

In an attempt to redress this loss of fighting power and reduce running costs, all six of these frigates were given an extensive half-life modernisation. This involved the replacement of obsolete weapons and sensors with modern systems found in newer ships, thereby achieving a higher degree of standardisation throughout the fleet. In view of today's high manning costs, this refit also involved the installation of numerous automatic labour saving devices in an attempt to reduce manning levels.

In late 1976, VAN SPEIJK paid off for her half-life refit, at Den Helder dockyard, which commenced early in 1977 and was completed just two years later on 3rd January 1979. However, the refits of VAN GALEN (15th July 1977 – 30th November 1979), VAN NES (31st March 1978 – 28th November 1980) and TJERK HIDDES (15th December 1978 – 7th August 1981) took two and a half years. The last two ships, EVERTSEN (18th July 1979 – 26th November 1982) and ISAAC SWEERS (1st July 1980 – 28th October 1983) were due to complete their refits in December 1981 and August 1982 respectively but they were delayed by about eight months due to lack of civillian labour in naval dockyards.

HNLMS VAN GALEN, 2/80 after modernisation   Mike Lennon

    The refits involved the removal of the 4.5 in gun and Mk 10 mortar and the installation of a single OTO-Melara 76 mm gun forward of the bridge. The flight deck was extended aft by plating over the mortar well and the hangar made telescopic to accommodate a Lynx helicopter equipped with dipping sonar. The ASW capability was enhanced by installation of two triple Mk 32 ASW torpedo tubes. These were mounted to port and starboard alongside the mainmast in VAN SPEIJK but in other ships they were relocated abreast the after end of the hangar. A dramatic increase in anti-surface capability was achieved by fitting canisters for Harpoon surface-to-surface missiles immediately aft of the funnel. Although intended to ship eight of these anti-ship missiles, budgetary constraints meant that in practice normally only two Harpoon canisters were carried. Unfortunately,

HNLMS VAN GALEN, 12/11/84. Note new IR suppression funnel cap   L. van Ginderen

it proved impossible within the limits of the budget and timescale available to replace the Seacat launchers with a more modern system and consequently, the weapons and associated radar/FCS were updated. The electronic suite was completely overhauled and the modern SEWACO-V data handling system installed, the operations room rebuilt, a DA 05 combined search aerial replaced the earlier SGR 105 while the old Mk 10 mortar-associated sonars were removed and replaced by more modern equipment including a compact VDS for deep sea use. All these new systems plus the LW 03 air surveillance radar on the mainmast, the M45 radar for the 76 mm gun and both M44 radars for the Seacat system were fully integrated with the newly installed data system. One most beneficial effect of these changes was the reduction of the ships' complement from 254 to 190 — a reduction of over 25% and a significant saving in costs. All ships were fitted eventually with new infra-red suppression funnel caps while EVERTSEN and ISAAC SWEERS were equipped with the US SQR18A towed array sonar with its winch on the port quarter.

HNLMS ISAAC SWEERS, entering Portsmouth 25/9/85, showing US SQR18A TASS fitted on the quarterdeck
L. van Ginderen

On 11th February 1986 the Dutch and Indonesian Governments signed a contract for the transfer of two (with an option for two more) Van Speijk class frigates to the Indonesian Navy. The ships were provided with all spare parts but not helicopters and the Dutch Navy was responsible for initial training of the Indonesian crews. Subsequently, TJERK HIDDES was transferred to Indonesia on 1st October 1986, being renamed AHMED YANI while her sister VAN SPEIJK was transferred one month later becoming SLAMET RIYADI. VAN GALEN became the Indonesian ship YOS SUDARSO on 2nd November 1987 and a year later was joined by VAN NES which was renamed OSWALD SIAHANN on 31st October.

In July 1989 it was announced that the Indonesians had obtained the last two frigates of the class at a cost of Df 179 milllion ($US 85 million) with EVERTSEN being transferred in November 1989 (becoming ABDUL HALIM PERDAMA KUSUMA) and her sister ISAAC SWEERS scheduled to follow twelve months later. These six Dutch-built Leander class frigates represent a substantial enhancement of the Indonesian Navy's capability both in terms of relative modernity and fleet homogeneity and presumably can be expected to remain in service until early next century.

The Indonesian frigate YOS SUDARSO (ex-HNLMS VAN GALEN) leaving Den Helder 17/12/87
L. van Ginderen

INDIAN LEANDERS

In the mid 1960s the Indian Navy started an ambitious project to construct its own major warships using a proven foreign design, in this case the British "Broad-beam" Leander general purpose frigate design. The decision followed a visit to India by Vickers Shipbuilding and a technical aid agreement between Vickers Shipbuilding/Yarrow & Co Ltd and the Indian Government. A total of six ships were constructed by Mazagon Docks Ltd, Bombay from October 1966 to July 1981 and, as might be expected from a fifteen year building span, there were considerable variations between the ships of the class. Although little known in the West, they are significant for being the first major warships built in Indian yards and show a most interesting design evolution.

The six ships were intended to be repeats of ANDROMEDA, displacing 2,450/2,962 tons with an overall length of 372 feet and a beam of 43 feet. As with their British cousins they are powered by two sets of geared steam turbines developing a total of 30,000 shp and driving two shafts to give a top speed of 28/29 knots. NILGIRI, which was the first of the class to be completed, was commissioned on 3rd June 1972 with an armament consisting of a twin 4.5 in gun turret forward, two 20 mm guns to port and starboard abreast the bridge, a quadruple launcher for Seacat missiles on the centreline of the hangar roof and a Mk 10 mortar in a well aft. Her electronic suite was entirely British consisting of a Type 965 with AKE-1 array long range radar aerial on the mainmast, together with a Type 993 target acquisition/surface search set and a Type 978 navigation radar on the foremast. A MRS3 fire control system is fitted on the bridge with a GWS22 Seacat director aft of the mainmast while the sonar outfit consists of Type 184 medium range search set and a Type 199 VDS set.

HIMGIRI, the second ship to be completed, commissioned on 23rd November 1974 with an identical armament to NILGIRI, but showed considerable differences in her electronics suite. Thus, a Dutch DA 08 long range radar replaced the ageing Type 965 while a modern ZW 06 target indicating aerial replaced the Type 993. Furthermore, while the GWS 22 Seacat control system was retained, a Dutch M45 system replaced the MRS3 on the bridge. Subsequent ships showed a further progressive increase in Dutch influence resulting in a hybrid Anglo-Dutch design of considerable capability with heavier armament, a full suite of Dutch radars and a bigger helicopter. Thus, the next pair of ships, UDAYGIRI and DUNAGIRI, which were commissioned on 1st February 1977 and 18th February 1976 respectively, were completed with two quadruple Seacat launchers and their associated Dutch M44 FCS on top of the hangar. In this pair of ships the Type 199 VDS gear was omitted although the VDS well aft was retained. NILGIRI, HIMGIRI, UDAYGIRI and DUNAGIRI were equipped to operate a French Alouette III helicopter which was too long for the original Leander hangar and so a telescoping hangar had to be fitted to accommodate this machine.

Indian Leander class frigate UDAYGIRI at Spithead 26/6/77   John G. Callis

The last pair, TARAGIRI and VINDHYAGIRI, which commissioned on 16th May 1980 and 8th July 1981 respectively, were completed to a greatly modified design. Thus, the Mk 10 mortar and VDS were removed completely to make enough space to operate a Westland Sea King helicopter with a telescopic hangar and the Canadian Bear-Trap haul-down gear. In these two ships an open deck has been left below the flight deck for handling the mooring gear while openings have been left in the hull sides beneath the helicopter deck at the stern. The shipboard ASW armament has been enhanced by the installation of two sets of triple AS torpedo tubes abreast the hangar at upper deck level and a twin Bofors 375 mm AS rocket launcher fitted forward of the 4.5 in gun turret. Finally, there are reports that VINDHYAGIRI and TARAGIRI are due to receive two SS-N-2 Soviet-made surface-to-surface missiles in lieu of the gun turret. All the various changes effected in this pair must have led to a significant increase in displacement and even caused topweight problems. There have also been reports that these two ships have been fitted with more powerful engines than those in their earlier sisters but no details have been released. All six vessels will be fitted with chaff launchers in due course and the earlier four ships will be equipped to the same standard as VINDHYAGIRI and TARAGIRI. In the spring of 1985, Westinghouse were awarded a £2,500,000 contract to construct five ASW transducer arrays for the Indian Navy. The arrays for hull-mounted and variable depth sonars will be installed on two of these Leanders in Bombay as part of an Indian Navy ASW upgrading programme.

Although only six Leander class frigates were built for the Indian Navy, their experience in blending together British and Dutch technology seems to have been instrumental in the production of an indigenous frigate design of great ingenuity. Thus, in the Godavari class, the basic Leander hull has been lengthened by 42.9 feet and broadened by 4.6 feet to allow a pair of Sea King ASW helicopters as well as four Soviet surface-to-surface missiles to be carried. The classic Leander hull shape has been modified and the raised forecastle replaced by a long bow with a slight sheer. The superstructure, foremast and funnel are clearly Leander in style but have been enlarged as have the hangar and flight deck area.

The three ships of the Godavari class were ordered in the late 1970s from Mazagon Docks Ltd, GODAVARI was laid down on 2nd June 1978, launched on 15th May 1980 and commissioned for service on 10th December 1983. GANGA, the second ship, was laid down in 1980, launched on 21st October 1981, and completed on 30th December 1985 while GOMATI laid down in 1981, was launched on 19th March 1984 and entered service on 16th April 1988.

As completed, GODAVARI is understood to have a displacement of 3,600 tons (4,100 full load) and measures 414.9 feet overall with a beam of 47.2 feet. Propulsion is by means of classic Leander machinery consisting of two sets of geared steam turbines which develop 30,000 shp and drive two shafts, each with five bladed propellers, to give a top speed of about 27 knots. Steam for her turbines is provided by two Babcock and Wilcox three-drum boilers and she has a range of 4,500 miles at 12 knots. By Western standards, her complement is relatively large for a ship of her size consisting of 51 officers and 262 men. However, by comparison with Western nations, India has a surfeit of cheap manpower of an age available for military service.

Her armament and electronic suite is a curious mixture of Soviet and Western technology quite unlike that seen in any other design. Her gun armament consists of a Soviet-made twin 57 mm gun mounting on the forecastle and four Soviet-made twin 30 mm Gatling guns, sited in pairs abreast the foremast, to provide ''last ditch'' defence against incoming anti-ship missiles. The armament is completed by four Soviet SS-N-2C Styx surface-to-surface missiles (in four single cells/launchers) situated abaft the twin 57 mm gun mounting, a twin launcher for Soviet SA-N-4 Gecko point-defence surface-to-air missiles mounted on the centreline forward of the bridge and two sets of triple torpedo tubes for A244 S ASW torpedoes.

GANGA at the Malaysian Fleet Review in 5/90   *Mike Louagie collection*

Her electronic outfit consists of a Soviet ''Head Net'' 3D long range air warning radar and a Dutch ZW 06 surface search/navigation radar on the foremast plus a Dutch DA 08 medium range air search aerial on the mainmast. The twin 57 mm gun mounting is controlled by a Soviet ''Muff Cob'' FCS situated on the centreline between the two pairs of SS-N-2C missiles while the SA-N-4 missiles are controlled by a Soviet ''Pop Group'' FCS on top of the bridge. Finally, the four twin 30 mm Gatling guns are controlled by a pair of ''Drum Tilt'' FCS fitted to port and starboard halfway up the mainmast. The rest of her electronic suite is something of a mystery but is reported to include a British Type 184 medium range sonar, a Selenia IPN-10 combat information data system plus a range of electronic warfare equipment.

Perhaps the most surprising feature of GODAVARI and her sisters is that they are able to accommodate and operate a pair of large Westland Sea King ASW helicopters in a hull which is understood to be 600 tons lighter, 12.9 feet shorter and 2.4 feet narrower than the highly successful Canadian Iroquois class destroyers! Assuming the reported dimensions and capability to be correct, it would appear that GODAVARI represents an effort to cram the proverbial ''quart into a pint pot'' and that rather too much may have been attempted on the displacement. This may account for the cancellation of a second trio of these ships. At the moment it is too early to judge whether or not this is a successful design especially as there is a certain amount of doubt about GODAVARI's exact displacement and dimensions. Therefore, although she is heavily armed and has a pleasant appearance, she will, for the moment, have to remain another of those intriguing mysteries of the East.

NEW ZEALAND LEANDERS

By the mid 1950s the Royal New Zealand Navy needed to replace its six slow and elderly Loch class frigates with new construction. Consequently in February 1957 the Royal New Zealand Navy ordered a pair of Rothesay class Type 12 frigates, OTAGO and TARANAKI which entered service in 1960 and 1961 respectively. This replacement programme was completed eventually by the construction of a pair of British-built Leander class frigates although, the latter were completed several years after the last of the Loch class vessels had been sold for scrap.

WAIKATO, the first of the pair, was ordered from Harland and Wolff, Belfast on 14th June 1963. Her future Commanding Officer, Captain S F Mercer, was based in London at the time and signed the instrument of contract and, when the ship was laid down on 10th January 1964, he welded her first seam. Subsequently, he was present at her launch on 18th February 1965 and commissioned her on 16th September the following year. She remained in British waters until the spring of 1967 and eventually arrived in New Zealand waters in May of that year. As completed WAIKATO was almost identical to a standard British Leander, except for the modified funnel with its paired extensions to the funnel cap. Subsequently, she completed a long refit and modernisation in July 1977 emerging from the dockyard with triple Mk 32 ASW torpedo tubes to port and starboard abaft the boats, no Mk 10 mortar, no VDS gear, and her hangar and flight deck enlarged to take a Lynx helicopter — the aircraft remains to be ordered and for the time being WAIKATO will have to make do with an obsolete Wasp.

CANTERBURY, the second New Zealand Leander was ordered from Yarrows of Scotstoun in August 1965 to the 'Broad beam' or Andromeda type. Subsequently, she was laid down on 12th April 1969 and launched just over a year later on 6th May. She was commissioned on 22nd October 1971, with the crew that had brought BLACKPOOL back to the United Kingdom earlier that year after five years service on loan to the Royal New Zealand Navy. As completed, CANTERBURY differed from an "as-designed" "Broad beam" Leander by the absence of the two single 20 mm guns, VDS gear and a Mk 10 mortar aft and the presence of paired USN pattern Mk 32 torpedo tubes abaft the boats as well as extensions to the funnel uptakes. After several months of trials

HMNZS CANTERBURY, 13/10/78   M. R. Dippy

in British waters CANTERBURY finally arrived in New Zealand in August 1972 just four years after being ordered. In May 1982 CANTERBURY relieved HMS AMAZON on station East of Suez so that the British frigate could be deployed to the South Atlantic. CANTERBURY was modernised during 1988-89 receiving a new RCA FCS for her 4.5 in guns, improvements to her radar and ESM, fitting of the SRBOC chaff launching system as well as having her flight deck extended.

In the early 1980s the Royal New Zealand Navy had a potentially serious problem of how to replace the rapidly ageing Type 12s OTAGO and TARANAKI which were over twenty years old and in need of replacement. Fortunately for New Zealand, the notorious British Defence White Paper of 1981 led to the abandonment of the planned Seawolf conversions of five Batch 3 Leanders and the early disposal of some of the older Batch 1 Ikara conversions. Consequently, it was announced in October 1981 that New Zealand would be acquiring the unmodernised Batch 3 frigate BACCHANTE and the older Ikara-armed DIDO. Following a brief refit BACCHANTE was transferred to New Zealand at Portsmouth on 1st October 1982 and formally commissioned as WELLINGTON three days later. After a few days of trials in the Portsmouth area, WELLINGTON left British waters

HMNZS WELLINGTON (ex-HMS BACCHANTE) at Portsmouth 11/10/82   L. van Ginderen

on 11 October, and on her arrival in Auckland was docked for a major refit. This involved the removal of the obsolete Mk 10 mortar as well as the VDS gear (which was retained as a spare for SOUTHLAND, ex DIDO) and the updating of weapons and sensors wherever possible. During this refit some steel plating was replaced between decks and the interior layout altered to incorporate the installation of additional fuel tanks to increase her range. On the upper deck, two sets of triple tubes for ASW torpedoes were installed and the mortar well plated over, thereby allowing the flight deck to be extended further aft. A US R-76C5 gunfire control system replaced the MRS3 fitted since completion and an entirely new weapon control system installed along with a general purpose computer. The frigate's surface search radar was upgraded while her radar intercept system was replaced and the Marconi NTC-1 communications system fitted. During this refit, WELLINGTON was fitted with the SRBOC chaff launching missile decoy system to replace the older Corvus rocket launchers which had been removed before the ship left the United Kingdom.

Work had begun on WELLINGTON in mid 1983 and the Royal New Zealand Navy had hoped originally to have the job completed in two years. Unfortunately, two factors combined to delay completion of work on this frigate. Firstly delays of up to six months were experienced with the delivery of over 90 per cent of the equipment ordered from overseas. Secondly, a 25 per cent shortage of civilian manpower in Auckland Naval Dockyard coupled with unscheduled work on operational warships and salvage operations on the sunken Greenpeace flagship RAINBOW WARRIOR meant that work on WELLINGTON was not completed until June 1986. Subsequently, on 7th July of that year, the frigate left the Auckland Naval Base at the start of an exhaustive series of post-refit trials. WELLINGTON's refit lasted three and a half years and cost $NZ 54 million and was the largest job ever undertaken by Auckland's Devonport Dockyard.

DIDO was handed over to New Zealand on 18th July 1983 at Portsmouth and renamed SOUTHLAND. Next day she sailed from Portsmouth to begin a refit at Vosper Thornycroft's yard at Southampton before being formally commissioned on 21st December 1983. While under refit, SOUTHLAND's machinery and weapons systems were extensively overhauled and her weapon fit improved by removing the obsolete Mk 10 mortar and replacing it with two sets of triple Mk 32 tubes for ASW torpedoes — the only Batch 1 Ikara armed Leander to be so equipped. After an extensive series of trials and work-up exercises in British waters, SOUTHLAND sailed for New Zealand on 9th May 1984 arriving at Wellington on 13th July. Subsequently, SOUTHLAND has been engaged in the usual round of exercises and courtesy calls and the latter included a visit to Shanghai, in company with her sister CANTERBURY, during the period 19th-22nd July 1987. SOUTHLAND was scheduled to receive a major refit in 1988 but this was cancelled because of cost and was reported that she was given a limited thirty five week-long refit starting in 1989.

HMNZS SOUTHLAND (ex-HMS DIDO), 2/2/88 arriving at Bluff, New Zealand. She was the only Leander class frigate to be equipped with Mk 32 triple torpedo tubes which were fitted abaft the hangar   P. Davey

The purchase of WELLINGTON (ex-BACCHANTE) and SOUTHLAND (ex-DIDO) from the Royal Navy has enabled the Royal New Zealand Navy to replace two older Type 12 frigates quickly and relatively cheaply as well as creating a four ship Leander squadron. While this is fine from a logistical point of view, it must not be forgotten that SOUTHLAND is twenty six years old and WAIKATO twenty three years old and that both vessels will need replacement within five to ten years. Furthermore, as WELLINGTON and CANTERBURY were completed in 1969 and 1971 respectively, and are already at least halfway through their service lives the New Zealand Navy is likely to be faced with the problem of block obsolescence of its frigate force before too long. In an attempt to overcome this problem, the New Zealand Government announced on 29th August 1989 that it was going to purchase two MEKO 200 ANZ Anzac class frigates with the option to purchase a further pair but these ships are not expected to enter service until 1995 at the earliest.

CHAPTER EIGHT

THE LEANDERS IN PERSPECTIVE

Numerous authors have claimed that the ships of the Leander class were built to one of the most successful of post Second World War frigate designs. While this may be true so far as the Royal Navy is concerned, is the statement valid when applied internationally? Twenty six of these frigates were constructed for the Royal Navy thereby becoming the Service's most numerous class of escorts built since 1945. Not surprisingly, this fact is often quoted in arguments attempting to prove that the Leander design is particularly successful. However, it could be that the Royal Navy had been kept so short of funds since 1945 that the Leander design was all that could be afforded. If this was the case then many commentators were making a virtue of necessity.

The success or failure of the Leander frigate design must be evaluated in terms of the requirements of the tasks that it was intended to perform as well as the ability of its contemporaries to fulfil the same roles. Furthermore because most warships remain in service for at least twenty years, another factor that must be taken into account is the ability of the design to remain viable in the face of technological developments. The pace of the latter is such that it is reasonable to expect any warship to become obsolescent within a decade of entering service and consequently, success or failure can be measured also in terms of the adaptability of the original design.

The procurement of a warship and its equipment is a complex and protracted process involving numerous steps. A key point to remember is that this process is one of evolution and therefore represents the development of existing concepts, technology, ships and their equipment. The ideal sequence of events starts with the definition of a need which is followed by feasibility studies leading to the generation of a Staff Requirement (N) from which specifications are derived. Thereafter, manufacturers are invited to tender for the work and ideally, the best proposal "fit for purpose" is selected. In practice, this means usually that either the cheapest or the politically most expedient proposal is chosen. For example, it is not unknown for orders for equipment to be placed deliberately so as to ensure continuity of employment for voters in particular Parliamentary constituencies. The next stage involves the construction of prototypes followed by a prolonged period of trials during which the suitability of the equipment is evaluated. However, prototypes for ships are constructed rarely, the best recent example being the minehunter WILTON which was built to test the feasibility of glass reinforced plastic in ship construction. The tests carried out at this stage will indicate what modifications are necessary to enable the equipment to fulfil its intended role(s). Alternatively, the trials may reveal so many shortcomings that the project has to be cancelled.

THE HISTORICAL CONTEXT

The period since the end of the Second World War has seen a continuous decline in the British economy and its manufacturing base, which has been paralleled by a progressive reduction in the size and capability of the British surface fleet. In 1945, the Royal Navy had a substantial force of battleships, aircraft carriers, cruisers and destroyers which could be combined together to form a powerful strike groups. In addition, there were also a large number of ships dedicated for use as convoy escorts. Since 1945, the Royal Navy's surface strike forces have been eliminated progressively as ships paid off for disposal without replacement when they reached the end of their service lives. This process has been such that it is possible to argue that the Type 42 destroyers, which can be considered as the successor to the County class fleet escorts, represent the last vestige of the once formidable British surface strike fleet.

Two other major factors acted to reduce the size of the United Kingdom's naval forces. First, there was the withdrawal from Empire, which was all but completed by 1968, and meant that there was no longer a need to maintain substantial naval forces east of Suez. Second, during the relevant period there has been the continuing NATO commitment to maintain considerable military and air forces in West Germany. These two factors together with a limited budget and the ever-rising cost of modern military hardware have caused a progressive reduction in the size of the fleet because it has proven impossible to replace existing ships on a one-for-one basis. This has resulted in the major role of the British surface fleet being reduced to little more than that of an escort force tasked with ensuring that, in times of crisis, American reinforcements could be transported to Europe in the event of a threat by the armies of the Warsaw Pact.

Consequently, the Royal Navy's escort forces have experienced a period of relative prosperity since 1945 with the construction of a range of specialist frigate designs as well as a class of light aircraft carriers. All of these ships were designed to defend trans-Atlantic convoys against attack by Soviet submarines and long range aircraft. It was assumed that when acting in this capacity, British warships would be operating under an "umbrella" provided by the Royal Air Force and United States Navy (USN) battle groups containing aircraft carriers.

As has been described in Chapter 1, in the period from 1945 to 1955 the Admiralty considered that high priority had to be given to the introduction into service of a range of escorts capable of countering the fast submarines then known to be under development. The end result was the construction of fifteen Type 12 and twelve Type 14 ASW frigates as well as the conversion of recently-built but obsolete fleet destroyers. Despite the high capability of such specialised frigate types when operating in their intended roles, it was soon recognised that to be effective these ships had to be brought together in flotillas composed of a mixture of dedicated ASW, AA and AD ships. Clearly, it would be impossible to fulfil such a criterion on every occasion during wartime with the result that flotilla effectiveness would be reduced while the vulnerability of the individual ships of a group would be increased. Another factor influencing frigate procurement in the mid 1950s was the recognition of the relatively high cost of operating a range of different ship classes, each with different equipment and operating procedures, compared to the cost of operating one class of ships capable of undertaking a range of different tasks. Such consideration gave rise to a Naval Staff Requirement resulting in the construction of seven second rate Type 81 General Purpose frigates of the Tribal class.

THE CHOICE

The Tribals, which were expected to supplement future British main-stream specialist escort vessel designs, were intended to be effective, if not outstanding, ships in three major roles, namely ASW, AA and AD. However, during 1958-60, and in the absence of Naval Staff Requirement, the existing Type 12 design was reworked to produce a general purpose frigate which fulfilled all the criteria that had to be met by the Type 81 design. The Admiralty was now faced with choosing between two competing designs.

Although the two designs were for ships of similar tonnage, equipment and armament, their hull forms, machinery and layout were quite different. For example, the Type 81 design featured the COSAG propulsion system while that for Leander was far more conservative utilising the same machinery as the existing Type 12. While both designs incorporated a landing platform and hangar for a Westland Wasp light helicopter, the facilities for helicopter maintenance, storage and operation were far superior in the Leander design although some helicopter pilots preferred the position of the flight deck in the Tribals. In terms of armament, the Type 81 would appear to have been superior to the Leander because it possessed two Seacat close range AA missile systems (designed for two twin Mk V 40 mm gun mountings) with good fields of fire. By comparison the Seacat launcher in the Leander design had a very restricted field of fire being fitted on the port side of the after end of the hangar. Similarly, on paper at least, the Tribal had the advantage of a higher rate of fire from its two single 4.5 in guns. However, the fully enclosed twin 4.5 in gun turret aboard the Leanders must have been safer and more efficient in rough weather. Both designs made use of the same range of radar, sonar and communication systems and, because of the incorporation of full air-conditioning, were capable of operating in climates ranging from the dry heat of the Persian Gulf, to the cold of the Arctic and the humidity of Singapore. However, because of their gas turbines, the Tribals could start almost immediately from "cold" without having to spend several hours raising steam.

The decisive factors in the adoption of the Leander general purpose frigate design therefore would appear to be hull form, sea keeping, speed and, of course, cost. The Tribals were still under construction as the Leander design was being finalised and consequently, there was no in-service information about the sea-keeping qualities of the Type 81 hull form. By comparison, the Leander design made use of the proven and outstanding Type 12 hull form which had been at sea in operational service since HMS TORQUAY commissioned for service on 10th May 1956. Another important factor may have been that because the Leander hull was somewhat larger than that of the Tribal, it gave the former greater potential for being updated to meet the demands posed by technological progress. Furthermore, the Leander design was for a ship capable of 29 to 30 knots whereas the maximum speed of the Type 81 was 26.5 knots. This factor may well have proved to be decisive for a navy which was beginning to experience a shortage of modern escorts capable of the 28 to 30 knots necessary to accompany and defend task groups built around aircraft carriers. Finally, with regard to cost, the Tribals would appear to have been somewhat more expensive to build with the first of class, ASHANTI, reportedly costing £5,220,000 whereas LEANDER, which was completed two years later, cost £4,810,000. A major factor in this cost difference was the "learning curve" phenomenon. Thus, it can be argued that ASHANTI was more expensive because she was the first ship to be built to an entirely new design, whereas LEANDER was yet another ship incorporating the main features of the Type 12 design.

Once selected, the Leander design set the pattern for British escort construction during the next decade with twenty six units being built while only seven Type 81 frigates were constructed. Furthermore, production of specialist frigates of the Type 12 Rothesay class as well as Type 41 AA and Type 61 AD frigates was abandoned.

OVERSEAS CONTEMPORARIES

While there can be little doubt that the Royal Navy adopted the best of the general purpose escort designs available in the United Kingdom, comparisons with equivalent contemporary overseas ships are difficult to achieve for a variety of reasons. First of all, during the period from 1945 to 1960, most of Britain's European NATO allies were concerned with rebuilding fleets that had been shattered during the Second World War. In such circumstances, construction of specialised escort vessels was likely to have been given a lower priority than in the United Kingdom, which still retained the elements of a relatively modern and powerful fleet, but was concerned with countering the next generation of fast submarines. Secondly, the United States Navy possessed an enormous reserve of recently-constructed, large, fast and well-armed destroyers which could be modified for service as ASW ships. Consequently, the main thrust of the United States Navy's ASW escort construction programme was aimed at the production of an updated and faster version of the wartime destroyer escort.

The French E50 class escort LE CORSE R. H. Osborne collection

FRANCE

In the immediate post-war period the French Navy set about the task of constructing a fleet to replace that destroyed during the 1939 — 45 war. One result was the completion of fourteen T47 Surcouf, five T53 Duperre class and one T56 La Galisonniere class destroyers between 1955 and 1962. These ships can be considered to be French equivalents of the Royal Navy's Daring class destroyers and so do not fit into the category of a general purpose frigate. In addition, the French also built four E50 Le Corse and fourteen E52 Le Normand class 1,250 ton, 28 knot, sea-going convoy escorts. Armed with three twin mountings for 57 mm AA guns, one sextuple 375 mm ASW mortar and four triple banks of ASW torpedo tubes, these ships were comparable to the Royal Navy's somewhat smaller and less capable Type 14 frigates. Finally, the nine 1,750 ton Commandant Riviere class ships completed for the French Navy between 1962 and 1965 were intended to serve as a colonial

COMMANDANT BORY, a dual-purpose colonial aviso/escort   R. H. Osborne collection

aviso in peacetime and as a NATO convoy escort in wartime. Although these diesel-powered 25-knot ships had been built to a dual purpose design, they did not possess the overall capability of a Leander class frigate because of their limited ASW capability and the inability to operate a helicopter.

NETHERLANDS

The Dutch Navy took delivery of four Holland class (2,211 tons, 32 knots) and eight Friesland class (2,497 tons, 36 knots) destroyers in 1954/55 and 1956/58 respectively. Designed specifically for ocean ASW operations, the hull form of these long-serving ships was reminiscent of comtemporary British destroyers and even incorporated side armour. Both classes were armed with two twin 4.7 in DP gun turrets and two 375 mm ASW mortars and were the first European destroyers to be completed without torpedoes. Compared to a Leander class frigate, these destroyers were larger, faster, had a heavier gun armament and an equally comprehensive array of radars but lacked a medium range search sonar and the capability of operating a helicopter. However, such comparisons may be misleading because the designs for these destroyers were finalised about ten years before that of the frigate.

The Dutch Friesland class destroyer ROTTERDAM at Portland 8/6/75   R. H. Osborne collection

Subsequently, in 1962 the Dutch Navy ordered the first four of a class of six frigates built to a modified version of the Royal Navy's Leander class general purpose frigate design. As has been described above, the Leander class was selected by the Dutch because, at that time, it was the best frigate design available at short notice.

ITALY

Post war Italian escort construction began with the completion during 1957-59 of four 1,807 ton Canopo class frigates which were capable of 26 knots, armed with four 3 in guns and made use of a considerable amount of American equipment. These were followed into service by four 25 knot ships of the Bergamini class which

CANOPO, 9/3/59   Aldo Fraccaroli

VIRGINIO FASAN, 17/3/71   *Aldo Fraccaroli*

were completed in 1961/62 and proved to be a considerable disappointment because too much was attempted on a displacement of just 1,410 tons. Both classes represent attempts to produce up-dated versions of the American wartime destroyer escort type and consequently fall into the same category as the USN's Dealey class and the French E50/E52 group.

The next pair of frigates, ALPINO and CARABINIERE, which were completed in 1968, displaced 2,000 tons and were equipped with a CODAG (COmbined Diesel And Gas turbine) propulsion system giving a maximum speed of 29 knots. As completed, ALPINO measured 371.8 feet overall with a beam of 43 feet and was armed with six single 76 mm guns, one ASW mortar and two sets of ASW torpedo tubes. Furthermore, they are capable of operating two ASW helicopters and have a comprehensive array of sensors including VDS. Clearly, although these two ships were comparable in size and AA/ASW capability to Leanders then under construction, the design of the latter was finalised at least three years before that of ALPINO. However, because of their appearance and propulsion system this pair of frigates should be considered to be contemporaries of the Royal Navy's Type 21 Amazon class. As might be expected with ships designed to serve predominantly in the Mediterranean, none of these Italian frigates had the sea-keeping ability of a Leander class frigate.

ALPINO, 1/5/69   *Aldo Fraccaroli*

WEST GERMANY

Two views of the West German frigate AUGSBURG of the "Köln" class at Antwerp, 16/3/87 L. van Ginderen

The first class of major warships to be built for the Bundesmarine were the six 2,100 ton frigates of the Köln class which were constructed between 1957 and 1964. Powered by bulky and rather inefficient CODAG machinery producing a maximum speed of 28 knots, KÖLN and her sisters were armed with two 100 mm and six 40 mm guns plus two four-barrelled ASW rocket launchers and four 21 in torpedo tubes. The hull featured low freeboard plus sharp flare with the end result that the ships were wet forward and prone to lively pitching motions. Furthermore, the ships which were cramped, complex and had no margin for modernisation, were obsolete within a few years of their completion. Despite their obvious limitations, KÖLN and her sisters remained important units of the Federal German Navy for over twenty years before being replaced by the six much larger and more capable frigates of the Bremen class

CANADA

The Canadian St Laurent class frigate HMCS ASSINIBOINE *R. H. Osborne collection*

Since the end of the Second World War, the Royal Canadian Navy has concentrated its efforts on anti-submarine warfare and during the period from 1950 to 1964 it constructed seven St. Laurent, seven Restigouche, four Mackenzie and two Annapolis class destroyer escorts. The four classes utilized the same hull form and machinery but showed differences in armament and equipment during the prolonged period of construction. The St Laurent class was designed by a team led by Rowland Baker RCNC during an eight year secondment to Canada. Despite obvious differences, his design met the same specification as the British Type 12 but incorporated various measures, such as plated masts, well rounded deck edges and a turtle back forecastle, to reduce problems of icing by permitting easy egress of water that came aboard. Operational service proved that the Type 12 and St Laurent class escorts and their derivatives were the two best seakeeping types of ships of their size in the NATO fleet.

USA

In 1945, the United States Navy possessed an enormous number of large, fast and well armed destroyers which were suitable for conversion into ASW escorts. Consequently, the first post war American destroyer escorts of the 1,314 ton Dealey class were intended to replace the 22 knot, war-built coastal PC boats. Thirteen

The Dealey class destroyer escort USS HOOPER, photographed in 1961   *L. van Ginderen*

ships of the Dealey class were built for the United States Navy during the period from 1952 to 1958 while the Norwegian and Portuguese Navies constructed a further five and three ships respectively to a modified form of the design. The Dealeys were considered to be expensive and consequently four 21 knot, 1,314 ton diesel-powered ocean escorts of the Claude Jones class were constructed between 1957 and 1960. This class was found to be totally inadequate for ASW warfare and was succeeded by the two ships of the far larger 1,882 ton Bronstein class which entered service in 1963.

This pair of ships became the progenitors of a long line of large ASW escorts which were completed for the United States Navy during the subsequent eleven years, but even by 1959 it had become clear that with a speed of only 26 knots the Bronsteins were going to be too slow to operate with ASW Task Forces. Consequently, it

The Garcia class frigate USS BRUMBY in 1984   *R. H. Osborne collection*

was decided to take advantage of new lightweight pressure-fired steam boilers to build a class of larger, faster and more capable escorts. The result was that ten Garcia class ASW frigates and six Brooke class ships armed with the Tartar area defence AA missile were constructed between 1962 and 1968. The ships of the Garcia class, which displace 2,441 tons (3,371 tons full load) with an overall length of 414 feet and a beam of 44 feet, are capable of at least 27 knots. Their armament consists of two single 5 in DP guns, the long range ASROC (Anti-Submarine ROCket) ASW system, two sets of triple tubes for ASW torpedoes and a DASH (Drone Anti-Submarine Helicopter). The latter, which was an unmanned equivalent of the Westland Wasp light helicopter used in the Royal Navy, was a total failure. In the Brooke class the after 5 in gun mounting was replaced by a

USS RAMSEY, a Brooke class frigate photographed from the aircraft carrier USS MIDWAY (CV41) in 3/87
L. van Ginderen

single-arm launcher for Tartar missiles. Both classes had a comprehensive range of radar, sonar and communications systems. In many ways, these ships are the American equivalent of the Leanders but it would appear that the design of the latter just pre-dates that of the Garcia and Brooke classes. However, in several respects, the smaller Leanders were superior because of their better seakeeping ability, the possession of a point defence AA weapon system and an effective ASW helicopter. In one respect, the Garcias were superior because they were equipped with the long range SQS-26 sonar but, this advantage was nullified because the Leanders possessed the low rpm machinery which was necessary for the installation of noise-reduced (NR) propellers. The latter were ten years ahead of the propellers in any other navy and consequently the Leanders were much quieter and more capable ASW ships than the Garcias.

ADAPTABILITY

The Leander design predated the introduction of long range AA and ASW missiles and it is apparent that the design was continued in production for some years after it had become obsolete. Unfortunately, the Naval Staff were divided with regard to the characteristics of the next generation of frigates hoping to obtain a smaller, cheaper and better ship. This was clearly impossible but the indecision resulted in the prolongation of an existing design rather than replacement by a new one. However, as has been described above, the Leander design proved to be very adaptable with the ships being capable of undergoing large scale reconstruction during which the latest weapons and sensors were fitted. Thus, eight ships of the class became specialist ASW ships when equipped with the Ikara AS missile while a further seven received Exocet missiles and a greatly enhanced AA armament. Finally, five ships were modified to incorporate the weapons and sensors being fitted in the succeeding and much larger Type 22 frigates. Clearly, the adaptability of LEANDER and her sisters must contribute to the overall conclusion that they belong to a successful design of frigates.

It is interesting to compare the capability of ARIADNE, the last Leander to be completed to a 13 year-old design, which entered service in February 1973 with that of AMAZON, the first of the gas turbine-powered Type 21 frigates, which was completed in May 1974. Thus, although the Type 21 as completed had a greater displacement and a higher maximum speed, there was no significant difference in the fighting capability of either ship. Indeed, it is possible to argue that ARIADNE was the superior ship because at that time she possessed

The Type 21 frigate HMS ANTELOPE, 30/6/75　　　　　　　　　　　　　　　　　　　　　　　*Vosper Thornycroft Ltd*

an on-board ASW weapon and was capable of fulfilling the AD role. Subsequently, the fire-power of the Amazon class frigates was increased by the addition of four Exocet anti-ship missiles and the STWS-1 ASW torpedo system. Unfortunately, the Type 21s lacked sufficient margin to permit the installation of the next generation of weapons and sensors and consequently have become increasingly obsolescent during the 1980s. By comparison, the Leander design proved to be able to absorb numerous technological changes with the result that the reputation of the class was not only maintained but actually enhanced. Finally, service in the South Atlantic showed that the Type 21s were fragile and prone to serious cracking which required extensive and expensive rectification.

HMS AMAZON 2/7/87 at Portsmouth. Note hull strengthening amidships and Exocet missiles forward of the bridge　　*L. van Ginderen*

FINAL CONCLUSIONS

It is difficult to argue that the Leanders were world-beaters in comparison with other navies' ships built or used for the same tasks because, as has been shown above, they have no real contemporaries other than the Royal Navy's Tribal class. As completed, the first two batches of Leanders were the best ships of their size anywhere in the world and featured the best sonars then available (Types 170 and 177), an action information organisation and a superb hull. Furthermore, their low rpm machinery and NR propellers made the Leanders the quietest and most effective ASW ships then in service while, the "learning curve" effect made them fairly cheap. Significantly, six Leander class frigates were built in Holland for the Netherlands Navy — the ultimate compliment.

The Leander design was intended to provide a general-purpose escort which was capable of meeting the challenges anticipated in the late 1950s. There is no doubt that when they entered service in the 1960s the first two batches of ships were fully capable of fulfilling their intended roles. By the late 1960s however the design was beginning to be overtaken by developments in missile, aircraft and submarine technology. A similar fate befell overseas warships designed at the same time as the Leanders. Consequently, those ships of the class completed after 1969 looked rather dated when compared with new construction for other navies. Unfortunately, lack of decision by the Staff resulted in the continued construction of by then obsolescent frigates so as to maintain the number of escorts available to the Royal Navy.

However, the Leanders were derived from a very good hull that proved capable of adaption into a general purpose frigate design and were suitable for considerable modernisation more than fifteen years later. Consequently, the Royal Navy was able to carry out far greater changes to armament and equipment in Leander class frigates than was the case with their overseas contemporaries. For example, the Batch 3 Seawolf conversion effected in five ships must rank as one of the most complete and successful reconstructions ever carried out on vessels of this size. The result was the conversion of five obsolete, gun-armed escorts into first rate ASW frigates with most of the capability of the succeeding Type 22 class.

It is however worth remembering the old adage that "reconstruction never pays" and consequently, the conversions effected in twenty ships of the Leander class may well be yet another manifestation of the continued weakness of the British economy since 1945. In this case, the Leander modernisation programme could be seen as clear evidence that the nation could not afford to order new ships to replace obsolescent tonnage. By comparison, the strength of the American economy has been such that the USN has been given the resources to order large number of modern replacement escorts without the need to reconstruct ageing ships. This proved to be the case for the sixteen frigates of the Garcia and Brooke classes which were never modernised throughout their careers with the USN and have now been transferred to several allied navies.

In conclusion, the Leander frigate design can be considered justifiably a success. The ships proved to be capable of fulfilling their intended tasks with economy as well having sufficient margin to permit useful modernisation and it is significant that their successors, the Type 22 Broadsword class, had to be approximately 1,600 tons heavier, 60 feet longer and at least 5.5 feet greater in the beam in order to accommodate all the weapons and sensors necessary in the 1980s for a seaworthy, first class ASW frigate. LEANDER and her sisters can therefore be regarded as representing the maximum that could be hoped for in a relatively small but ocean-going hull and consequently their high reputation would seem to be well deserved.

The Batch 1 Type 22 frigate HMS BRILLIANT at Greenwich 7/12/84   T. Bolton

BRITISH LEANDER CLASS BUILDING PROGRAMME

Estimates	Name	No	Laid Down	Launched	Commissioned	Deck Letters
1959-60	LEANDER	F109	10.4.59	28.6.61	26.3.63	LE
	AJAX	F114	19.10.59	16.8.62	11.12.63	AJ
	DIDO	F104	2.12.59	22.12.61	18.9.63	DO
	PENELOPE	F127	14.3.61	17.8.62	31.10.63	PN
	AURORA	F10	1.6.61	28.11.62	9.4.64	AU
	EURYALUS	F15	2.11.61	6.6.63	16.9.64	EU
	GALATEA	F18	29.12.61	23.5.63	25.4.64	GA
1961-62	ARETHUSA	F38	17.9.62	5.11.63	24.11.65	AR
	NAIAD	F39	30.10.62	4.11.63	15.3.65	NA
	CLEOPATRA	F28	19.6.63	25.3.64	1.3.66	CP
1962-63	PHOEBE	F42	3.6.63	8.7.64	15.4.66	PB
	MINERVA	F45	25.7.63	19.12.64	14.5.66	MV
	SIRIUS	F40	9.8.63	22.9.64	15.6.66	SS
1963-64	JUNO	F52	16.7.64	24.11.65	18.7.67	JO
	ARGONAUT	F56	27.11.64	8.2.66	5.9.67	AT
	DANAE	F47	16.1.65	21.10.65	10.10.67	DN
1964-65	HERMIONE	F58	10.12.65	26.4.67	11.7.69	HM
	ANDROMEDA	F57	25.6.66	24.5.67	2.12.68	AM
	JUPITER	F60	3.10.66	4.9.67	13.8.69	JP
1965-66	BACCHANTE	F69	27.10.66	29.2.68	5.12.69	BC
	CHARYBDIS	F75	27.1.67	28.2.68	6.6.69	CY
	SCYLLA	F71	17.5.67	8.8.68	14.2.70	SC
1966-67	ACHILLES	F12	1.12.67	21.11.68	11.9.70	AC
	DIOMEDE	F16	30.1.68	15.4.69	2.4.71	DM
1967-68	APOLLO	F70	21.11.69	15.10.70	10.6.72	AP
	ARIADNE	F72	1.5.70	10.9.71	2.3.73	AE

BRITISH LEANDER CLASS: CONVERSIONS

BATCH 1 IKARA GROUP
Conversion cost £7,600,000-£23,000,000

Name	No	Yard	Dates	Notes
LEANDER	F109	Devonport	6.70 - 1.73	
AJAX	F114	Devonport	10.70 - 2.74	Harbour Training Ship 8.85
GALATEA	F18	Devonport	10.71 - 9.74	
NAIAD	F39	Devonport	1.73 - 6.75	
EURYALUS	F15	Devonport	5.73 - 3.76	
AURORA	F10	Chatham	12.74 - 2.76	
ARETHUSA	F38	Portsmouth	10.73 - 4.77	Towed Array, Devonport 5.85-3.86
DIDO	F104	Devonport	7.75 - 10.78	

Transferred New Zealand 18.7.83

BATCH 2 EXOCET GROUP
Conversion cost £13,800,000-£47,700,000

Name	No	Yard	Dates	Notes
CLEOPATRA	F28	Devonport	7.73 - 12.75	Towed Array, Devonport 1982-3
PHOEBE	F42	Devonport	8.74 - 4.77	Towed Array, Chatham 1982 & Devonport 1983
SIRIUS	F40	Devonport	3.75 - 2.78	Towed Array, Devonport 1981-3
MINERVA	F45	Chatham	12.75 - 4.79	
ARGONAUT	F56	Devonport	2.76 - 3.80	Towed Array, Devonport 1982-3
DANAE	F47	Devonport	8.77 - 4.81	
PENELOPE	F127	Devonport	1.78 - 1.82	

BATCH 2 NAVIGATION TRAINING SHIP

Name	No	Yard	Dates
JUNO	F52	Rosyth	.83 - 2.85

BATCH 3 SEAWOLF GROUP
Conversion cost £60,000,000-£79,700,000

Name	No	Yard	Dates	Notes
ANDROMEDA	F57	Devonport	3.78 - 2.81	
CHARYBDIS	F75	Devonport	6.79 - 7.82	
JUPITER	F60	Devonport	1.80 - 10.83	
HERMIONE	F58	Chatham	1.80 - 6.83	Completed Devonport 7.83-1.84
SCYLLA	F71	Devonport	11.80 - 12.84	Conversion suspended 4-6.82

FOREIGN LEANDERS AND DERIVATIVES

Name	No.	Laid Down	Launched	In Service
AUSTRALIA (River class)				
SWAN	F50	18.8.65	16.12.67	20.1.70
TORRENS	F53	18.8.65	28.9.68	19.1.71
CHILE				
ALMIRANTE CONDELL	06	5.6.71	12.6.72	21.12.73
ALMIRANTE LYNCH	07	6.12.71	6.12.72	25.5.74
NETHERLANDS (Van Speijk class)				
VAN SPEIJK	F802	1.10.63	5.3.62	14.2.67
VAN GALEN	F803	25.7.63	19.6.65	1.3.67
TJERK HIDDES	F804	1.6.64	17.12.65	16.8.67
VAN NES	F805	25.7.63	26.3.66	9.8.67
ISAAC SWEERS	F814	5.5.65	10.3.67	15.5.68
EVERTSEN	F815	6.7.65	18.6.66	21.12.67
NEW ZEALAND				
WAIKATO	F55	10.1.64	18.2.65	16.9.66
CANTERBURY	F421	12.4.69	6.5.70	22.10.71
SOUTHLAND (ex-DIDO)	F104	2.12.59	22.12.61	18.9.63
WELLINGTON (ex-BACCHANTE)	F69	27.10.66	29.2.68	5.12.69
INDIA (Leander class)				
NILGIRI	F33	.10.66	23.10.68	3.6.72
HIMGIRI	F34	.67	6.5.70	23.11.74
UDAYGIRI	F35	1.73	9.3.74	1.2.77
DUNAGIRI	F36	14.9.70	24.10.72	18.2.76
VINDHYAGIRI	F38	.75	12.11.77	8.7.81
TARAGIRI	F41	.74	25.10.76	16.5.80
INDIA (Godavari class)				
GODAVARI	F20	2.6.78	15.5.80	10.12.83
GANGA	F22	.80	21.10.81	30.12.85
GOMATI	F23	.81	19.3.84	16.10.88

PARTICULARS OF BRITISH LEANDER CLASS FRIGATES

Displacement (Tonnages: standard/full load):
Batches 1 & 2, 2,380/2,860;
Batch 3, 2,500/2,962;
Batch 1 Ikara conversion, c.2,500/c.3,000; Batch 2 Exocet conversion, 2,700/3,200;
Batch 3 Seawolf conversion 2,790/3,300 tons.

Dimensions:	Batches 1 & 2: 372 (oa), 360 (pp/wl) x 41 x 18 feet (over props); Batch 3: 372 (oa), 360 (pp/wl) x 43 x 19 feet.
Machinery:	Two Babcock & Wilcox boilers operating at 550 psi and 850°F; English Electric geared steam turbines developing 30,000 shp and driving two shafts, each with a five-bladed propeller, at 220 rpm. Astern power 10,000 shp.
Max Speed:	28 knots; maximum trial speed 29.5 k; max speed on one boiler 24 k.
Oil fuel:	450 (Batches 1 & 2)/ 460 (Batch 3) tons.
Range:	4,000 nm at 15 knots; 1,000 nm at max speed.
Complement:	263 (Batches 1 & 2)/ 260 (Batch 3) (19 officers, 58 senior and 183 junior ratings); 250 (Ikara conversion); 223 (Exocet conversion); 214 (Seawolf conversion).
Aircraft:	One Westland Wasp. One Westland Lynx helicopter in Batch 2 & 3 conversions.
Armament as built:	Two 4.5 in DP guns in a Mk 6 mounting and two single Bofors 40 mm AA guns, one Mk 10 AS mortar, (first seven only); two 4.5 in DP guns in a Mk 6 mounting and two single 20 mm Oerlikon guns, one quadruple Seacat launcher, one Mk 10 AS mortar.
Batch 1 conversion:	One Ikara AS missile system, two quadruple launchers for Seacat SAM system, two single Bofors 40 mm AA guns, one Mk 10 AS mortar.
Batch 2 conversion:	Four Exocet MM38 surface-to-surface missiles, three quadruple launchers for Seacat SAM, two single Bofors 40 mm guns and two STWS-1 triple tubes for ASW torpedoes.
Batch 2A conversions:	Four Exocet MM38 surface-to-surface missiles, two quadruple launchers for Seacat SAM, two single Oerlikons 20 mm guns and two STWS-1 triple tubes for ASW torpedoes.
Batch 2 conversion — JUNO:	Two single 20 mm guns and two STWS-1 triple tubes for ASW torpedoes.
Batch 3 conversion:	Four Exocet MM38 surface-to-surface missiles, one sextuple launcher for Seawolf SAM, two single Oerlikon 20 mm guns, two STWS-1 triple tubes for ASW torpedoes.
Radar as built:	Navigation — Type 974 or 978; target indicating — Type 993 long range air-warning — Type 965; fire control — Type 903; IFF — Type 1010.
Batch 1 conversion:	Navigation — Type 1006; target indicating — Type 993 (later 994); fire control — Type 903; IFF — Type 1010.
Batch 2 conversion:	Navigation — Type 1006; target indicating — Type 994; long range air-warning — Type 965; fire control — Type 903; IFF — Type 1010.
Batch 2A conversion:	Navigation — Type 1006; target indicating — Type 994; fire control — Type 903; IFF — Type 1010.
Batch 2 conversion — JUNO:	Navigation — Type 1006; target indicating — Type 994; IFF — Type 1010.
Batch 3 conversion:	Navigation — Type 1006; target indicating/warning — Types 967/968 with IFF; fire control — Type 910.
Sonar as built:	Types 184, 177, 170, 182 199 (some ships), 162 & 185.
Batch 2 conversion:	Types 184M and 182.
Batch 2A conversion:	Types 2031 'I', 184M and 182.
Batch 3 conversion:	Types 2016, 162M and 182.

MACHINERY CONTRACTS — BRITISH BUILT SHIPS

Names	Builder (Yard No)	Machinery *
ACHILLES	Yarrow (2365)	White
AJAX	Cammell Laird (1285)	Cammell Laird
ANDROMEDA	HM DY Portsmouth	White
APOLLO	Yarrow (1002)	White
ARETHUSA	White (2015)	White
ARGONAUT	Hawthorn Leslie (758)	White
ARIADNE	Yarrow (1003)	White
AURORA	John Brown (721)	John Brown
BACCHANTE	Vickers, Tyne (189)	Vickers, Barrow
CHARYBDIS	Harland & Wolff (1675)	White
CLEOPATRA	HM DY Devonport	Cammell Laird
DANAE	HM DY Devonport	White
DIDO	Yarrow (2155)	Yarrow
DIOMEDE	Yarrow (2366)	Yarrow
EURYALUS	Scotts (691)	Scotts
GALATEA	Swan Hunter (1935)	Wallsend Slipway
HERMIONE	Stephen (697)	White
JUNO	Thornycroft (4207)	Thornycroft
JUPITER	Yarrow (2289)	Yarrow
LEANDER	Harland & Wolff (1591)	Harland & Wolff
MINERVA	V-A, Tyne (179)	Vickers, Barrow
NAIAD	Yarrow (2224)	Yarrow
PENELOPE	V-A, Tyne (165)	Vickers, Barrow
PHOEBE	Stephen (681)	Stephen
SCYLLA	HM DY Devonport	White
SIRIUS	HM DY Portsmouth	White
CANTERBURY	Yarrow (1001)	White
WAIKATO	Harland & Wolff (1657)	Harland & Wolff
ALMIRANTE LYNCH	Yarrow (1007)	Yarrow
ALMIRANTE CONDELL	Yarrow (1006)	Yarrow

* Some builders sub-contracted parts of the machinery

BRITISH LEANDER CLASS: MACHINERY AND VDS FIT

Name	Machinery	VDS Well	VDS Fitted
LEANDER	Y 100	yes	yes
AJAX	Y 100	yes	not originally
DIDO	Y 100	yes	yes
PENELOPE	Y 100	yes	yes
AURORA	Y 100	yes	yes
EURYALUS	Y 100	yes	not originally
GALATEA	Y 100	yes	not originally
ARETHUSA	Y 100	yes	yes
NAIAD	Y 100	yes	yes
CLEOPATRA	Y 100	yes	yes
PHOEBE	Y 136	yes	yes
MINERVA	Y 136	yes	no; well soon plated
SIRIUS	Y 136	yes	no; well soon plated
JUNO	Y 136	yes	no; well soon plated
ARGONAUT	Y 136	yes	no
DANAE	Y 136	yes	no
HERMIONE	Y 160	yes	yes
ANDROMEDA	Y 160	yes	yes; soon removed
JUPITER	Y 160	yes	yes
BACCHANTE	Y 160	yes	yes
CHARYBDIS	Y 160	yes	yes
SCYLLA	Y 160	yes	no; well soon plated
ACHILLES	Y 160	yes	no; well soon plated
DIOMEDE	Y 160	no; well plated before completion	
APOLLO	Y 160	no	no
ARIADNE	Y 160	no	no

BRITISH LEANDER CLASS: WEAPONS AND SENSORS

GUNS

4.5 in DP Twin mounting Mk 6
   Mk 5 guns in a Mk 6* Mod 3 semi-automatic, remotely controlled mounting, introduced in 1946 and no longer in production. Manufactured by Vickers Ltd, the turret weighs 45 tons; muzzle velocity is 2,400 ft/sec and maximum range against surface targets 24,000 yds, (normal 18,000 yds) and air targets 8,000 yds. Rate of fire 16 rounds per minute per barrel; maximum elevation 80°. The shell and cartridge are separately hoisted and manually loaded; projectile weight is 56 lb, cartridge 30 lb, cordite $12\frac{1}{2}$ lb. About 400 rounds per gun carried; gun crew 20 men.

Oerlikon 20 mm Mk 2 gun on Mk 7 single mounting
   Obsolete in 1945 but brought back into service as a close-range surface weapon. Locally controlled, the rate of fire is 480 rpm in maximum bursts of 60 rounds. The 70 cal gun weighs about 64 kg, total mount 560 kg, and the projectile 0.12 kg, total round weight about 0.24 kg. The range of the gun is about 5 km, but effective AA range c.1000 m. Muzzle velocity 835 m/sec; elevation +80° to -15°. Crew 2.

Bofors 40 mm/60 cal Mk 9 mounting
   Introduced in 1942, now obsolete but remains in service. This is an anti-aircraft and anti-surface (junk-basher) weapon with a maximum elevation of 80°. Rate of fire 120 rpm; maximum range 10 km but effective AA range of 2 km. Muzzle velocity 830 m/sec and projectile weight 0.9 kg, round 2.2 kg. Mount weight 1.75 tons. Crew 3.

GAM-B01 20 mm single mounting
   Manufactured by B-MARC under licence from Oerlikon-Buhrle of Zurich. Introduced to the Fleet in 1982 after the lessons learnt during the Falklands campaign. The weapon is belt-fed, gas-operated and intended principally for use against soft-skinned surface and air targets. The rate of fire is 1,000 rpm using a variety of ammunition eg. HE, incendiary and armour-piercing. Elevation from +60° to -15°; effective range is 2,000 m with a muzzle velocity of between 1,050-1,150 m/sec. Mount weight 0.5 tonne; projectile weight 0.13 kg, round 0.34 kg.

MISSILES

Seacat
   The missile is a close-range surface-to-air weapon fired from a 3-ton weight quadruple launcher. On firing, the missile is radio-command guided and can be tracked visually while being controlled by the aimer, or may be radar or TV directed using the associated fire-control director. To assist tracking one of the fixed tail fins carries a flare. The wings on the missile body are cruciform, swept back and pivot to control the direction of flight; they are mounted at 45° to the tail-fins. Seacat is 4.8 ft long, c. 7.5 in in diameter with a wing span of c. 26 in. Launch weight is 140 lb range 4,500 yds. A two-stage solid propellant motor powers the missile which is fitted with a HE warhead and either contact or proximity fuzes. Manufactured by Shorts of Belfast, who have recently updated the weapon.

Ikara — GWS 41
   Australian designed missile-body to carry a US Mk 44 or Mk 46 AS torpedo. Due to go to sea first in the Royal Navy aboard BRISTOL, the weapon underwent sea-trials on the newly converted LEANDER in early 1973. The missile is stowed in the magazine with the torpedo already fitted. When ready for use the fins are fitted in the missile handling room before the weapon moves on to the launcher, which has a fixed angle or 55°, and is mounted behind a curved screen or zareba. Guidance may be from the firing ship's own sensors or those of other ships in the vicinity. The range is 14 miles, missile length 3.42 m, wing span 1.50 m. About 16 rounds can be carried.

Exocet MM38 — GWS 50
   Manufactured by Aerospatiale in France, the missiles are fired from container-launchers fitted on a ramp. On firing the missile gains altitude, acquires the target with its own radar and then drops to within 2 or 3 metres of the surface to attack the target. Missile length is 5.12 m, diameter 34.4 cm, wing span 100 cm. The launch weight is 730 kg (warhead 160 kg), the two-stage solid-propellant motor giving a speed of just under Mach 1. Range is about 38 km.

Seawolf — GWS 25
   Developed by British Aerospace, Seawolf is a short-range, supersonic, anti-missile missile fired from a hand-loaded, six-barrelled launcher. The missile has a length of 2 m, 70 cm wing span and a diameter of 30 cm. Launch weight of the missile is 82 kg; the total weight of the fully loaded launcher is 3,500 kg. Guidance is via the Marconi Type 910 pulse-doppler radar, and target acquisition and designation by the combined Type 967/968 radar. Range is about 5 km.

CHAFF LAUNCHERS

Corvus
    This is an eight-barrelled launcher introduced from 1968 onwards. Fitted either side of the superstructure, the launcher fires rockets to dispense chaff or infra-red decoys intended to seduce incoming missiles away from the ship. Manufactured by Vickers, the original Knebworth rockets have now been replaced by the more sophisticated BBC rounds.

Mk 36 SRBOC
    The American Super Rapid-Blooming Offboard Chaff system was introduced in the Royal Navy in 1982 during the Falklands conflict. Manufactured by the American Hycor Corporation, the Mk 137 launchers are six-tubed and fire Mk 182 chaff-dispensing cartridges to provide clouds of chaff up to 244 m above the ship.

Wallop Barricade
    The Barricade systems are a family of lightweight, low-cost decoy systems capable of providing protection against both air and surface launched anti-ship missiles. Ammunition for the Barricade systems is available with preset delays to give payload delivery ranges between 800 and 1,800 metres for the 57 mm rockets.

TORPEDOES

US Mk 44 AS torpedo
    An anti-submarine torpedo, fired from either STWS-1 tubes, as part of the Ikara weapon, or from a helicopter. Now obsolete, the torpedo has an active-homing head and is powered by a battery and electric motor giving a speed of c.30 k. Length is 2.56 m, diameter 32.4 cm and weight 233 kg. Maximum operating depth is 300 m. Replaced in service by the Mk 46.

US Mk 46 AS torpedo
    Of the same dimensions as the earlier weapon, the Mk 46 employs liquid-fuel propulsion and is capable of multiple re-attack using active- or passive-homing; carries a 40 kg warhead. Deployed as the Mk 44, but is faster, deeper diving and has a greater endurance than.

Stingray
    British-built torpedo with active/passive acoustic homing to replace the above. Operating depths from 18 m to over 300 m. The weapon has a length of 2.597 m, diameter of 32.4 cm and a weight of 266 kg. Propulsion is by seawater-activated batteries powering an electric pump-jet propulsor which gives a speed of 45k; range 7000 m.

ANTI-SUBMARINE WEAPONS

Mk 10 AS Mortar
    The Mk 10 mortar was introduced in 1955, following trials with a prototype known as Limbo, in the Type 15 frigates ROCKET and RELENTLESS in 1951. Though the name Limbo has been attached to this weapon since its introduction, the production version is correctly known as the Mk 10 mortar. This is a three-barrelled weapon, pitch-and roll-stabilised and fires 394 lb projectiles (207 lb explosives) in salvoes forward, or to either side, of the ship. The barrels are 12 in. in diameter and are loaded pneumatically with the fuzes set automatically to a maximum depth of 1,200 ft. An uprated version of the Mk 10 mortar was fitted in the Ikara conversions. Range between 400 and 1,000 yds; 2 salvoes/minute.

AIRCRAFT

Westland Wasp HAS 1
    This helicopter was first introduced into the Fleet in 1963/64 and was withdrawn on 31 March 1988. Originally carried in all Leander class frigates in RN service, the Wasp was replaced by the modern Lynx in all but the Ikara conversions, JUNO and the unmodernised Batch 3 ships, leaving the latter without a helicopter since March 1988. The aircraft was powered by a Rolls Royce Nimbus engine which gave a cruising speed of 96 k. Range was about 230 nm, considerably less than the Lynx. The aircraft had a length of 12.29 m, rotor diameter of 9.82 m and a height of 3.6 m; the all-up weight was 2,457 kg. The Wasp could carry up two Mk 44 torpedoes, or a Mk 11 depth charge, or two AS-12 Nord anti-ship missiles. This 76 kg, 1.87 m long missile is wire-guided and has a range of 8,000 m.

Westland Lynx
 The Lynx is a larger, more powerful helicopter than the Wasp and is allocated to the Batch 2 Exocet conversions (with the exception of CLEOPATRA) and the Batch 3 Seawolf Leanders. Powered by two Rolls Royce Gem 60 turboshaft engines, the Lynx can be used in either anti-surface or anti-submarine roles depending on the weapon fit. In the former role the aircraft can carry up to four Sea Skua missiles. As an anti-submarine helicopter the Lynx can carry a maximum of two Mk 44, Mk 46 or Stingray ASW torpedoes or two depth charges. The aircraft is given an all-weather capability by the Seaspray 360° radar fitted in the nose. The maximum all-up weight of the Lynx is 5,126 kg, length 15.24 m and width with rotors turning 12.8 m. Operational range is about 430 nm with an endurance of 4 hours. When on deck the helicopter is secured using the Harpoon deck-lock.

RADAR

Type 903
 Fire-control set for the MRS3 director associated with the 4.5 in gun mounting. A modified version — Type 904 — is fitted in the Seacat director. The radar is based on the US Mk 35 and the MRS3 on the US Mk 56 director.

Type 904
 Fire control radar for GWS 22 Seacat SAM system, derived from Type 903.

Type 910
 Tracking radar for Seawolf, manufactured by Marconi.

Type 965
 This Marconi-built radar is a long-range air search set identified by the large AKE-1 antenna fitted to all UK Leanders as completed. The one ton antenna measures 26 ft x 8 ft x 6 ft and rotates at either 10 or 8 rpm, depending on the power supply. This P-band radar has been removed from the Ikara and Seawolf conversions, JUNO and some of the Exocet conversions. In these vessels the Type 965 antenna has been replaced by the Type 1010 Mk 10 IFF aerial. Later versions of the radar include moving target indication. Range over 200 nm.

Type 967/968
 These two antennae are mounted back-to-back and are used in conjunction with the Seawolf system. The Type 967 is an L-band pulse-doppler air search radar, which together with the S-band Type 968 surface search radar is linked to the Type 910 Seawolf tracking radar. The fully stabilised aerial rotates at 30 rpm. Marconi manufactured.

Type 974
 A high definition surface search set for navigation and anti-submarine detection. Operated on X-band. The aerial rotated at 24 rpm. Replaced in the 1960s by Type 978.

Type 978
 This is an X-band replacement for the Type 974, employing a "double-cheese" antenna. Based on the Decca 45 set, the ATZ antenna rotates at 24 rpm.

Type 993
 An S-band target indicating set, the Type 993 employs the familiar "quarter-cheese" antenna seen at the foremast head of many post-war British frigates and destroyers. The radar has been replaced in recent years by the Type 994. Range about 60 miles.

Type 994
 Based on the Plessey AWS-4 radar, the Type 994 has an improved performance over the Type 993 which it replaced, though the earlier antenna is still used. Installed from 1978 onwards.

Type 1006
 A Kelvin Hughes manufactured system which has largely replaced the Type 978 navigation radar in British frigates. Uses X-band and the antenna rotates at 24 rpm.

Type 1010
 IFF set, manufactured by Cossor. Commonly mounted on top of the AKE-1 antenna for Type 965, or separately on converted ships.

SONAR

Type 162
A bottom-classification set first introduced in 1948, The Type 162 detects objects using the shadow-effect. It employs three transducers, one mounted centrally on the keel forward and one each side at an angle of 25°. Solid-state version designated Type 162M. Still in service.

Type 170
This set was the first real post-war sonar to be fitted. Developed as a short range search and attack set for the Mk 10 AS mortar, the system underwent trials in 1949. The transducer is pitch- and roll-stabilised and electronically trained. The sonar is hull-mounted forward and has a searchlight beam type scan.

Type 177
This is a medium range, low-frequency sonar. The two-ton, hull-mounted transducer consists of four elements which produce a 40° fan-shaped beam. A pulse system is employed to give detection in the widest range of sea-conditions. Entered service in 1956 following trials in BROCKLESBY in 1954.

Type 182
The RN towed decoy for torpedoes.

Type 184/184M
A low-frequency sonar developed as a self-protection set for large ships and medium range search for destroyers/frigates, but replaced the Type 177. Some replaced by the solid-state version Type 184M which underwent trials in PENELOPE and was fitted as ships entered major refit.

Type 189
Cavitation indicator and noise level indicator.

Type 199
The variable depth sonar developed in Canada and first fitted in the Leander class. Could be lowered to a depth of about 250 ft and towed at 24 knots. Never proved very reliable in RN service and has now been withdrawn altogether.

Type 2016
Fitted to the Seawolf converted ships, this is a hull-mounted 360° scan passive search and attack set introduced to Fleet in 1979. Manufactured by Plessey.

Type 2031 'I'
Towed array sonar employing passive hydrophones at the end of a cable steamed from the ship. The trials version was evaluated in LOWESTOFT in the late 1970s. The production set was built to Admiralty Research Establishment, Portland, specifications. Fitted in PHOEBE, SIRIUS, ARGONAUT and ARETHUSA following trials on CLEOPATRA.

HMS CHARYBDIS, 7/90. Note "Cherry-Bee" on the funnel.   Michael Cassar

SOURCES

The major records, journals and books consulted in the preparation of this volume were:

Periodicals/Papers

R. N. Andrew & A. R. J. M. Lloyd
The Naval Architect (1981), No 1, pp 1-31
"Full Scale Comparative Measurements of the Behaviour of Two Frigates in Severe Head Seas"

J. S. Canham
The Naval Architect (1975), No 2, pp 61-94
"Resistance, Propulsion and Wake Tests with HMS PENELOPE"

R. H. Osborne
Warships Supplement (to Marine News) (1978), *52*, pp 10-15
"The Third Cod War 11/75-6/76"

R. H. Osborne
Marine News (1982), *36*, pp 19-21
"Dutch Leander Class Frigates"

R. H. Osborne
Ships Monthly (1982), *17*, No 4, pp 16-20; No 5, pp 28-30; No 6, pp 18-20
"From Bay to Broadsword"

R. H. Osborne
Marine News (1985), *39*, pp 624-628
"Additions and Alterations Effected in British Warships Since the Outbreak of the Falklands War in April 1982"

M. K. Purvis
The Naval Architect (1974), No 4, 189-222
"Post War Royal Navy Frigate and Guided Missile Destroyer Design 1944-69"

D. A. Sowdon
Marine News (1989), *43*, pp 451-452
"Garage Refit of HMS JUNO"

D. F. Whitwam & A. J. Watty
The Naval Architect (1979), No 2, pp 37-51
"Modernising the Leander Class Frigates"
This article also contains M. K. Purvis' comments on the lack of a Naval Staff Requirement for the Leander class frigates.

P. L. Young
Navy International (1984), *89*, pp 91-99
"The Royal Australian Navy"

Official Publications

"The United Kingdom Defence Programme: The Way Forward", HMSO, June 1981.
"The Falklands Campaign: The Lessons", HMSO, December 1982.
"Statement on the Defence Estimates 1984", HMSO, 1984.

Books

"A Century of Naval Construction", D. K. Brown, Conway Maritime Press 1983.
"An Engineer of Sorts", Cdr P du Cane, Nautical Publishing Co, 1971.
"Conways All the World's Fighting Ships 1947-82, Parts I & II", Conway Maritime Press, 1983.
"Falklands — The Air War", by R. A. Burden, M. I. Draper, D. A. Rough, C. R. Smith & D. L. Wilton, Arms and Armour Press, 1986.
"Janes Pocket Book of Naval Armament", edited by D. Archer, Macdonald Janes, 1976.
"Modern Warship Design and Development", N. Friedman, Conway Maritime Press, 1979.
"Naval Radar", N. Friedman, Conway Maritime Press, 1981.
"Sea Combat off the Falklands", A. Preston, Collins/Willow, 1982.
"Seek and Strike", W. Hackmann, HMSO Books, 1984.
"The Decline of British Sea Power", D. Wettern, Janes, 1982.
"The Royal Navy and the Falklands War", J. D. Brown, Leo Cooper, 1987.

SHIP AND CLASS INDEX

Ship names and photograph references are shown in *italics*

Abdul Halim Perdama Kusuma (ex-Evertsen),
  Indonesia 90
Achilles 42-4, *45,* 47-51, 52, 59, 82
Active 52
Agincourt 16
Ahmed Yahni (ex-Tjerk Hiddes), Indonesia 90
Aisne 16
Ajax 22, 32-3, 36, *38,* 46, 57-60, *61,* 65
Alert 28
Almirante Condell, Chile 69, *87*
Almirante Latorre, Chile 87
Almirante Lynch, Chile 69, *87*
Alnwick Castle 10
Alpino, ITS *102*
Amazon (A class) 19
Amazon (Type 21) 94, *107*
Annan *11*
Annapolis class, Canada 104
Andromeda 37-8, 42-4, 46, *47,* 48,
                                                        59, 79-80, *81,* 82, 91
Antelope *107*
Apollo 44-5, 48-9, *50,* 51-3, 82
Arethusa 33, 35, *36,* 37-8, 46, 56-8, 61, *62,* 65, 116
Argonaut 36, 38, 42, 46, 65-6, 70, 75, 76, *78,* 116
Ariadne 31, 44-5, 48-9, *50,* 51-2, 58, 69, 82, 107
Ashanti 27, 98
Assiniboine, HMCS *104*
Augsburg, FGS *103*
Aurora 32-3, 36, 38, 46, *56-7,* 58-9, 61-2, *63,* 64-5
Bacchante 38, 42-5, 47, *48,* 49, 82, 95
Baldur, Icelandic gunboat 47
Barrosa 16
Battle class, Early/1942 9
Battle class, Later/1943 ....9, 15, 16, 54
Bay class 9
Bergamini, ITS *102*
Bergamini class, Italy 102
Blackpool *20,* 94
Black Swan class 9
Blake 59
Bremen class, W Germany 102
Brilliant 83, *108*
Bristol 43, 52, 54, 113
Britannia 60
British Esk, mv 48
Broadsword 9, 79
Broadsword class 9, 108
Brocklesby *34,* 116
Bronstein class, USA 105
Brooke class, USA 105-6, 108
Brumby, USS *105*
Bustler, RMAS 64
CA class destroyers 14
Canopo class, Italy 101
Canopo, ITS *101*
Canterbury, HMNZS 47, *94,* 96
Caprice 14
Carabiniere, ITS *102*
Castle class 9
Cavalier 14
CH class destroyers 14

Charles F Adams class 55
Charybdis 38, 42, *43,* 48, 82, 116
Chichester *18*
Claude Jones class, USA 105
Cleopatra 33, 36, *37,* 38, 40, 42, 46, *66,* 68-9, 72,
                                                                75-6, 78, 115-6
CO class destroyers 14
Columbia, HMCS 37
Commandant Bory, FS *100*
Commandant Riviere class, France 100
Concord *14*
Corunna 16
County class 30, 97
Coventry 32
Crescent, HMCS 34
Crusader, HMCS 34
Danae 36, 38, 42, 46, 66, *67-8,* 71
Daring class 9, 99
Dealey class, USA 102, 105
Derwent, HMAS 85
Devonshire 54
Dido 22, 32-3, 36, 38, 44-5, *46,* 57-8, *59,* 95-6
Diligence, RFA 60
Diomede 42, 44-8, *49,* 50-1, 59, 82
Dunagiri, INS 92
E50 Le Corse class, France 99, 102
E52 Le Normand class, France 99, 102
Eagle 57
Eastbourne *19,* 20
Eilat, Israel 42, 46, 66
Ennerdale, RFA 46
Euryalus 32, 35-6, 38, 46, 52-3, 58, 61-2, *64,* 65
Evertsen, HNLMS *88,* 90
Falmouth 59
Fife 53
Flower class 9
Fort Victoria, RFA 43
Fowey 22, 32
Foxhound, RMAS 64
Friesland class, Netherlands 100
Frigate, 1945 Design 12-13
G class destroyers 9
Galatea 32, 35-6, 38, 46-7, 57, *58,* 62-3, 65
Ganga, INS 92, *93*
Garcia class, USA 105-6, 108
Godavari, INS 92-3
Godavari class, India 92
Gomati, INS 92
Gota Lejon, Chile 87
Grenville *13*
Gurkha *26-8,* 30-1
Hammerburg, USS 37
Hastings (i) 22
Hastings (ii) 22, 32
Hermione 30-1, 37-8, 42, *43,* 48, *80,* 82, *83*
Himgiri, INS 92
Holland class, Netherlands 100
Hooper, USS *105*
Hul-Vul (ex-Naiad) *64*
Hunt class 9
Intrepid 52

Isaac Sweers, HNLMS	88, *90*
Juno	36, 38, 42, 44, 47-8, *49,* 52, *53,* 66, 114
Jupiter	37-8, 42, 45, 48, *79,* 80, 82-3
Köln, FGS	103
Köln class, W Germany	103
Leander	*2, 7,* 22, *29,* 31-3, 35-8, 45-7, 54, *55,* 57, *58-9,* 61, *62-3,* 66, 87, 98, 107-8, 113, *120*
Leander class design	9, 29 et seq
Le Corse, FS	*99*
Lincoln	18
Liverpool	70
Loch class	9, 94
Loch Achray	*11*
Londonderry	22
Lowestoft	*22-5,* 52, 62, 75, 116
Mackenzie class, Canada	104
MATCH concept	25, 57
McCloy, USS	*105*
Meko 200 ANZAC class, Australia/New Zealand	96
Mermaid	*10*
Minerva	33, 36, 38, *41,* 46, 66, 69, *70,* 72, 76
Naiad	33, *35,* 37-8, 44, 46-7, 58-9, *60,* 61, 64
Nautilus, USS	20
Nilgiri, INS	91-2
Norfolk	66
O class destroyers	*9,* 19
Olympic Alliance, mv	*47*
Oswald Siahann (ex-Van Nes), Indonesia	90
Otago, HMNZS	22, 94-5
Overijssel, HNLMS	37
Parramatta, HMAS	*85*
PC class, USA	*105*
Pellew	*21*
Penelope	*32,* 33, 36-8, *39,* 40, 45-6, 48, 66, 71, *72-3,* 116
Phoebe	33, 36, 38, *40,* 42, *44,* 46, 66, 68-9, *74,* 75, *76,* 77-8, 116
Preserver, HMCS	72
Puget Sound USS	72
Puma	*17*
Rainbow Warrior, mv	95
Ramsey USS	*106*
Relentless	13, 65, 114
Restigouche class, Canada	104
Rhyl	22
River class	9
River class, Australia	54, 85
Robust, RMAS	52, 61, 64
Rollicker, RMAS	*63-4*
Rothesay class	29, 57, 85, 98
Rothesay	*22*
Rotterdam, HNLMS	*100*
St Brides Bay	*12*
St Laurent class, Canada	104
Salisbury	18, 60
Scarborough	20
Scorpion	*15*
Scylla	38-9, 42-3, 45, 47-8, 53, 80, 82, *83,* 84
Shamsher (ex Diomede), PNS	*51*
Sirius	33, 36, 38, *41,* 46, 66, 69, 75, *77-8,* 116
Slamet Riyadi (ex-Van Speijk), Indonesia	90
Southland (ex-Dido), HMNZS	60, 95, *96*
Stuart, HMAS	85
Surprise	28
Swan, HMAS	85, *86*
T47 Surcouf class, France	*99*
T53 Duperre class, France	*99*
T56 La Galisonniere class, France	*99*
Taragiri, INS	92
Taranaki, HMNZS	94-5
Teazer	14
Tenby	*20*
Tjerk Hiddes, HNLMS	88, 90
Torquay	48, 98
Torrens, HMAS	85, *86*
Tribal class	27, 29-30, 98, 108
Type 12 Frigate	9, 12, 19-22, 85, 94, 96, 98, 104
Type 14 Frigate	21, 98, 100
Type 15 Frigate, Full Conversion	13, 14
Type 16 Frigate, Limited Conversion	13, 14
Type 18 Frigate	13
Type 21 Frigate	102
Type 22 Frigate	48, 77, 79, 82, 84, 108
Type 23 Frigate	77
Type 41 Frigate	12, 17, 19-20
Type 42 Destroyer	97
Type 61 Frigate	12, 17, 19-20, 32, 98
Type 62 Frigate	14
Type 81 Frigate	27, 29, 78
Type XXI U-boats	12
Tyr, Icelandic gunboat	47
Udaygiri, INS	47, *92*
Undaunted	66
Van Galen, HNLMS	88, *89,* 90
Van Nes HNLMS	44, 88, 90
Van Speijk, HNLMS	88-90
Ver, Icelandic gunboat	47
Vigilant	*11*
Vindhyagiri, INS	7, 92
Virginio Fasan, ITS	*102*
Waikato, HMNZS	94, 96
Warspite	59
Weapon class	9, 15
Wellington (ex-Bacchante), HMNZS	*49, 95,* 96
Weymouth	22, 32
Whitby	19
Wilton	97
Yarmouth	22, 70
Yarra, HMAS	85
York	19
Yos Sudarso (ex-Van Galen), Indonesia	90
Z/C class destroyers	9
Zulfiqar (ex-Apollo), PNS	*51*
Zulu	27

R.I.P.

HMS LEANDER, 9/89, being expended as a target during NATO exercise "Sharp Spear" *Crown Copyright*